Analysing Economic Data

Palgrave Texts in Econometrics

General Editors: **Kerry Patterson** and **Terence C. Mills**

Titles include:

Simon P. Burke and John Hunter
MODELLING NON-STATIONARY TIME SERIES

Michael P. Clements
EVALUATING ECONOMETRIC FORECASTS OF ECONOMIC AND FINANCIAL VARIABLES

Lesley Godfrey
BOOTSTRAP TESTS FOR REGRESSION MODELS

Terence C. Mills
MODELLING TRENDS AND CYCLES IN ECONOMIC TIME SERIES

Terence C. Mills
ANALYSING ECONOMIC DATA

Kerry Patterson
A PRIMER FOR UNIT ROOT TESTING

Kerry Patterson
UNIT ROOTS TESTS IN TIME SERIES VOLUME 1
Key Concepts and Problems

Kerry Patterson
UNIT ROOTS TESTS IN TIME SERIES VOLUME 2
Extensions and Developments

Palgrave Texts in Econometrics
Series Standing Order ISBN 978-1-4039-0172-9 (hardback)
978-1-4039-0173-6 (paperback)
(*outside North America only*)

You can receive future titles in this series as they are published by placing a standing order. Please contact your bookseller or, in case of difficulty, write to us at the address below with your name and address, the title of the series and one of the ISBNs quoted above.

Customer Services Department, Macmillan Distribution Ltd, Houndmills, Basingstoke, Hampshire RG21 6XS, England

Analysing Economic Data

A Concise Introduction

Terence C. Mills

Department of Economics, Loughborough University, UK

First published 2014 by
PALGRAVE MACMILLAN

Palgrave Macmillan in the UK is an imprint of Macmillan Publishers Limited, registered in England, company number 785998, of Houndmills, Basingstoke, Hampshire RG21 6XS.

Palgrave Macmillan in the US is a division of St Martin's Press LLC, 175 Fifth Avenue, New York, NY 10010.

Palgrave Macmillan is the global academic imprint of the above companies and has companies and representatives throughout the world.

Palgrave® and Macmillan® are registered trademarks in the United States, the United Kingdom, Europe and other countries.

ISBN: 978-1-137-40189-2

A catalogue record for this book is available from the British Library.

A catalog record for this book is available from the Library of Congress.

Transferred to Digital Printing in 2013

Contents

List of Tables

List of Figures

1
Introduction

Abstract: *The aims, objectives and structure of the book are set out. The level of mathematics required is stated, and it is emphasised that key algebraic proofs are relegated to the end-notes of each chapter, where key references are also to be found. The provision in the accompanying website of both the data used in examples and of* Econometric Views *exercises for replicating the examples is discussed. A brief word on the notation used for cross-referencing is also provided.*

1.1 About the book

After nearly 40 years lecturing, teaching and examining various courses in quantitative techniques, probability, statistics and econometrics in departments of economics and business schools in the UK, I thought that the time was right to attempt to write a text book on analysing economic data that I would wish to use for an introductory course on the subject. This book is the result and, as a consequence, contains material that I think any student of economics and finance should be acquainted with if they are to have any understanding of a real, functioning economy rather than having just a working knowledge of a set of academically constructed models of some abstract aspects of an artificial economy.[1]

After an introductory chapter on basic descriptive statistics, attention is then given to issues such as how growth rates and index numbers might be computed, how moving averages can be used to decompose time series into various components such as the trend and seasonal, and to the various ways of deflating nominal measures to obtain real measures. The core techniques of correlation and regression are then introduced, again as descriptive methods of assessing the strength or otherwise of associations between sets of data, as well as some key ideas in regression that are essential to understanding the behaviour of economic variables. The basic framework of statistical inference is then developed from the fundamentals of probability and sampling, thus enabling inferences to be drawn from correlation and regression results. This leads naturally to a discussion of classical linear regression, where the role of inference and hypothesis testing is placed centrepiece, reflecting my view that with modern computer software the computational aspects of regression analysis should be relatively downgraded and the interpretational aspects emphasised: if economic theory gives us anything it is hypotheses that require detailed examination and testing. The regression model is then extended to an analysis of the various breakdowns of the classical assumptions that pervade the analysis of economic data and essentially make up the subject of econometrics. A final chapter introduces some basic time series concepts and models that are finding increased importance in economics and finance.

1.2 Mathematical level, focus and empirical exercises

Basic algebra and calculus are all that is required by way of mathematical background. Several proofs are relegated to the end-notes of each chapter: these notes also include key references, historical perspective and suggestions for further reading that should provide both a wider and a deeper understanding of the concepts. Throughout the book, many examples are used to illustrate methods and techniques. These typically involve actual data, often from the UK, as this is where my experience and interests lie. The content is thus suitable for undergraduate and some postgraduate students of economics, finance and related subjects as an initial foray into handling economic data from an integrated economic and statistical perspective. It will also provide a statistical foundation, if such is required, for any of the *Palgrave Texts in Econometrics.*

Empirical exercises accompany most chapters. These are based on the software package *Econometric Views* (or *EViews*), now the industry standard for econometric software, and illustrate how all the examples used in the book may be calculated and how they may be extended. The data is available in *EViews* workfiles available for download at http://www.palgrave.com/economics/millsaed/index.asp. It is assumed that readers already have a basic working knowledge of *EViews* or are prepared to obtain this knowledge via the extensive online help facility accompanying the package.[2]

A brief word on notation: as can be seen, chapter sections are denoted **x·y**, where **x** is the chapter and **y** is the section. This enables the latter to be cross-referenced as §**x·y**.

Notes

1 Many of the articles contained in Diane Coyle (editor), *What's the Use of Economics? Teaching the Dismal Science after the Crisis* (London Publishing Partnership, 2012) provide a similar perspective to the views offered here. In particular, Andrew Lo's statements that 'economists wish to explain 99% of all observable phenomena using three simple laws, like the physicists do, but ... have to settle, instead, for ninety-nine laws that explain only 3%' and that economists should 'place much greater emphasis on empirical verification of theoretical predictions and show much less attachment to theories rejected by the data' (Chapter 7, 'What post-crisis

changes does the economics discipline need? Beware of theory envy', p. 42) both resonate.

So, too, do John Kay's comments concerning deductive and inductive reasoning. 'Deductive reasoning of any kind necessarily draws on mathematics and formal logic; inductive reasoning is based on experience and above all careful observation .... Much scientific progress has been inductive: empirical regularities are observed in advance of any clear understanding of the mechanisms that give rise to them. ... Economists who assert that the only valid prescriptions in economic policy are logical deductions from complete axiomatic systems [nevertheless] take prescriptions from doctors who often know little more about these medicines than that they appear to treat the disease' (Chapter 8, 'The map is not the territory: an essay on the state of economics', p. 53).

The importance of good data analysis to the 'real' world of economics and society is illustrated using many contemporary examples by Michael Blastland and Andrew Dilnot, *The Tiger that Isn't* (London, Profile Books, 2007). The theme of high quality data analysis linked with innovative economic theorising is the basis of Steven D. Levitt and Stephen J. Dubner, *Freakonomics: A Rogue Economist Explores the Hidden Side of Everything* (London, Allen Lane, 2005) and its sequel, *Superfreakonomics* (London, Allen Lane, 2009). Tim Harford's *The Logic of Life* (London, Little Brown, 2008) is written in a similar vein, and also listen to his and Blastland's BBC Radio 4 programme *More or Less*.

The importance of statistical analysis to medicine and health is the subject of Stephen Senn's *Dicing with Death: Chance, Risk and Health* (Cambridge University Press, 2003), while an excellent debunking of many popular myths and fears using statistical analysis is Dan Gardener, *Risk: The Science of Politics and Fear* (London, Virgin Books, 2008). This theme is also taken up by John Brignell in *Sorry, Wrong Number! The Abuse of Measurement* (Brignell Associates, 2000) and *The Epidemiologists: Have They Got Scares for You!* (Brignell Associates, 2004). Two thought-provoking books, particularly in light of the financial crisis of the late 2000s, are those by Nassir Nicholas Taleb, *Fooled by Randomness: The Hidden Role of Chance in the Markets and in Life* (London, Texere, 2001) and *The Black Swan: The Impact of the Highly Improbable* (London, Penguin, 2007), although these require an appreciation of the concepts of probability and probability distributions, which are discussed in Chapters 7–11.

2 *EViews 7* (Version 7.1) is used throughout: see *EViews 7* (Quantitative Micro Software, LLC, Irvine CA: www.eviews.com).

2
Presenting and Summarising Data

Abstract: *The distinction between cross-sectional and time series data is made, and examples of world income inequality and inflation in the UK are introduced to illustrate this distinction. Summary statistics for location, dispersion and asymmetry are developed, along with a composite pictorial representation of these measures, known as a boxplot. Scatterplots to graphically represent bivariate relationships between economic variables are introduced by way of an example investigating the long-run relationship between UK inflation and interest rates.*

2.1 Data types

Economic data is either *cross-sectional* or arrives in the form of *time series* (although it is possible to combine the two types to form a *panel data set*, which must be analysed using methods that are too advanced to be covered in this book).[1] While both types often lend themselves to standard forms of statistical modelling, they each have unique features that require their own special techniques of data analysis.

Cross-sectional data: income inequality

Table 2.1 shows a standard measure of income – purchasing power parity (PPP) adjusted real per capita gross domestic product (GDP, measured in US dollars) – for a wide cross-section of countries in 2009, and thus provides information on relative global living standards.[2]

Even though the data has been listed in ascending order of income, and while such an *enumeration* provides all the information that is available, it is still quite difficult to get any overall 'feel' for certain crucial features of the data, such as the *average* income and the extent of *deviations* about that average, which would give some indication of income inequality across the world.

Figure 2.1 encapsulates the complete data set in the form of a *histogram*, which is a means of representing the underlying *frequency distribution* of the data.

Here the income values have been classified into groups (known as *classes*) of width \$2500: thus the left-hand *bar* shows that there are 52 countries with incomes in the range \$0 to \$2500; the next bar shows there are 26 countries with incomes in the range \$2500 to \$5000; etc. The two extreme values have been identified (Luxembourg (\$84,572) and Qatar (\$159,144): extreme values are often known as *outliers*) and the positions of the UK (\$33,410) and the US (\$41,147) have also been identified out of interest.

An important feature of cross-sectional data is that, typically, the ordering of the data is irrelevant; Table 2.1 could just as well have listed the countries in descending order of income, or even alphabetically, rather than in ascending order.

Time series data: inflation in the UK

Table 2.2 lists the rate of UK inflation, as a percentage, for every year from 1751 up to 2011.[3]

TABLE 2.1 *Real per capita GDP, in PPP adjusted dollars, for 2009*

1	Zimbabwe	143	47	Congo, Republic of	2223
2	Congo, Democratic Republic of	231	48	Kyrgyzstan	2300
3	Burundi	368	49	Pakistan	2353
4	Liberia	397	50	Uzbekistan	2384
5	Somalia	461	51	Yemen	2401
6	Niger	534	52	Moldova	2496
7	Eritrea	593	53	Laos	2636
8	Central African Republic	647	54	Papua New Guinea	2753
9	Malawi	652	55	Philippines	2839
10	Ethiopia	684	56	Vietnam	2871
11	Togo	733	57	Mongolia	3170
12	Madagascar	753	58	India	3238
13	Mozambique	759	59	Morocco	3292
14	Guinea-Bissau	818	60	Micronesia, Federal States	3329
15	Guinea	823	61	Swaziland	3444
16	Sierra Leone	871	62	Honduras	3608
17	Burkina Faso	900	63	Paraguay	3702
18	Comoros	915	64	Cape Verde	3770
19	Mali	999	65	Bolivia	3792
20	Rwanda	1030	66	Syria	3995
21	Benin	1116	67	Sri Lanka	4034
22	Uganda	1152	68	Indonesia	4074
23	Timor-Leste	1155	69	Kiribati	4092
24	Afghanistan	1171	70	Fiji	4284
25	Tanzania	1189	71	Guyana	4336
26	Kenya	1205	72	Maldives	4461
27	Nepal	1209	73	Bhutan	4566
28	Ghana	1241	74	Jordan	4646
29	Chad	1276	75	Iraq	4709
30	Lesotho	1309	76	Namibia	4737
31	Cote d'Ivoire	1344	77	Angola	4756
32	Bangladesh	1397	78	Egypt	4957
33	Haiti	1444	79	Georgia	5063
34	Gambia	1464	80	Armenia	5376
35	Senegal	1492	81	Algeria	6074
36	Mauritania	1578	82	Ecuador	6171
37	Sao Tome and Principe	1681	83	Guatemala	6288
38	Zambia	1765	84	Tunisia	6300
39	Cambodia	1768	85	El Salvador	6341
40	Cameroon	1807	86	Ukraine	6415
41	Tajikistan	1873	87	Vanuatu	6531
42	Solomon Islands	2004	88	Samoa	6547
43	Nigeria	2034	89	Dominica	6580
44	Djibouti	2061	90	Albania	6641
45	Sudan	2188	91	Turkmenistan	6936
46	Nicaragua	2190	92	Marshall Islands	7092

Continued

TABLE 2.1 *Continued*

93	Bosnia and Herzegovina	7117	142	Libya	19233
94	Peru	7279	143	Portugal	19904
95	Montenegro	7318	144	Slovak Republic	19986
96	St. Vincent & Grenadines	7378	145	Oman	20541
97	China	7431	146	Saudi Arabia	21542
98	Colombia	7528	147	Malta	21668
99	South Africa	7587	148	Equatorial Guinea	22008
100	Macedonia	7682	149	Barbados	22928
101	Thailand	7799	150	Czech Republic	23059
102	Tonga	7862	151	Bahrain	23538
103	Belize	8444	152	Puerto Rica	23664
104	Serbia	8532	153	Seychelles	23864
105	Jamaica	8801	154	Slovenia	24956
106	Botswana	8868	155	South Korea	25048
107	Venezuela	9123	156	Israel	25559
108	Brazil	9356	157	Greece	27305
109	Mauritius	9487	158	Spain	27647
110	Azerbaijan	9619	159	Italy	27709
111	Romania	9742	160	New Zealand	27878
112	Dominican Republic	9919	161	Bahamas	28382
113	Turkey	9920	162	Taiwan	28716
114	Panama	10187	163	France	30837
115	Gabon	10276	164	Trinidad & Tobago	31057
116	Iran	10620	165	Japan	31980
117	Suriname	10644	166	Finland	32186
118	Bulgaria	10923	167	Germany	32492
119	Uruguay	11067	168	Ireland	33406
120	Costa Rica	11227	169	United Kingdom	33410
121	Malaysia	11309	170	Denmark	33929
122	Cuba	11518	171	Belgium	34625
123	Mexico	11634	172	Sweden	35246
124	Kazakhstan	11733	173	Canada	36234
125	Argentina	11960	174	Hong Kong	36293
126	Chile	12007	175	Iceland	37212
127	Grenada	12024	176	Austria	37413
128	St. Kitts & Nevis	12755	177	Switzerland	39632
129	Latvia	12777	178	Netherlands	40574
130	Belarus	12782	179	United States	41147
131	Lebanon	12907	180	Australia	41288
132	St. Lucia	13079	181	Brunei	46206
133	Lithuania	14189	182	Kuwait	46747
134	Russia	14645	183	Singapore	47313
135	Palau	14988	184	Norway	49974
136	Antigua & Barbuda	15047	185	Macao	51111
137	Croatia	15084	186	Bermuda	52091
138	Estonia	16294	187	United Arab Emirates	53855
139	Poland	16376	188	Luxembourg	84572
140	Hungary	16521	189	Qatar	159144
141	Cyprus	18998			

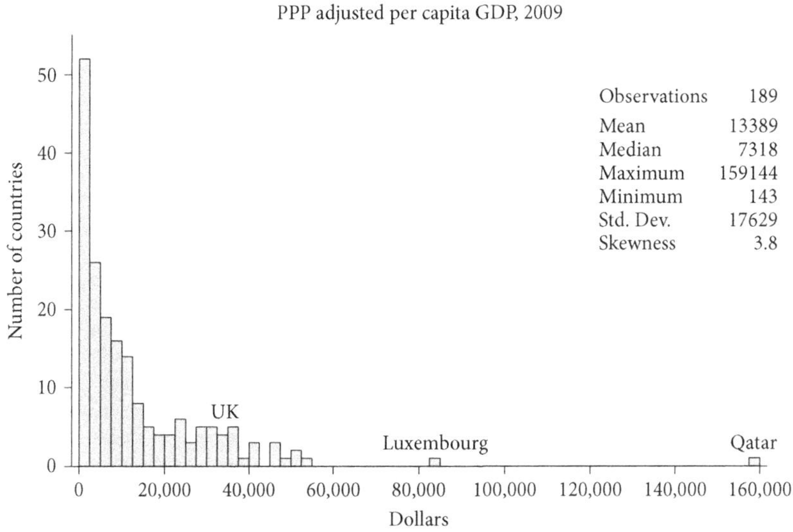

FIGURE 2.1 *Histogram of the per capita GDP data shown in Table 2.1*

We may thus interpret the data as a time series of UK inflation in which, unlike the previous example, the ordering of the data *is* paramount: it would make no sense to list the data in, for example, ascending order as it would lose the 'calendar characteristics' of inflation entirely.

Figure 2.2 thus presents a time series plot of inflation, whose features we shall discuss in detail in subsequent examples.

Constructing a histogram is still valid, and often useful, for a time series, and this is shown in Figure 2.3.

Here the data has been classified into classes of width 2.5%, and we see that there are two outliers: a maximum of 36% in 1800 and a minimum of −23% two years later in 1802; the Napoleonic wars obviously caused great volatility in the British economy. The general tendency, however, is for inflation to cluster in the range 0 to 5%.

2.2 Summary statistics: measures of location (central tendency)

Various *summary* statistics also accompany the two histograms. The *mean* is defined in the following way. If there are N data values, which are denoted $x_1, x_2, \ldots, x_N$, then we define the (*sample*) *mean*, $\bar{x}$ (read as 'x-bar'), as

TABLE 2.2 *UK inflation, % per annum, 1751–2011*

1751	−1.96	1795	10.59	1839	6.86	1883	−1.06	1927	−2.70	1971	9.44
1752	4.00	1796	6.38	1840	1.83	1884	−2.15	1928	0.00	1972	7.13
1753	−1.92	1797	−10.00	1841	−1.80	1885	−3.30	1929	−1.11	1973	9.10
1754	3.92	1798	−2.22	1842	−8.26	1886	−1.14	1930	−2.81	1974	16.04
1755	−5.66	1799	12.50	1843	−11.00	1887	−1.15	1931	−4.05	1975	24.24
1756	4.00	1800	36.36	1844	0.00	1888	1.16	1932	−2.41	1976	16.54
1757	21.15	1801	11.85	1845	4.49	1889	1.15	1933	−2.47	1977	15.85
1758	0.00	1802	−23.18	1846	4.30	1890	0.00	1934	0.00	1978	8.30
1759	−7.94	1803	−5.17	1847	12.37	1891	1.14	1935	0.63	1979	13.39
1760	−3.45	1804	2.73	1848	−12.84	1892	0.00	1936	0.63	1980	17.99
1761	−5.36	1805	15.93	1849	−6.32	1893	−1.12	1937	3.75	1981	11.87
1762	3.77	1806	−3.82	1850	−5.62	1894	−1.14	1938	1.20	1982	8.61
1763	3.64	1807	−2.38	1851	−3.57	1895	−1.15	1939	2.98	1983	4.59
1764	8.77	1808	4.07	1852	0.00	1896	−1.16	1940	16.76	1984	4.98
1765	3.23	1809	9.38	1853	9.88	1897	2.35	1941	10.89	1985	6.08
1766	1.56	1810	2.86	1854	14.61	1898	0.00	1942	7.14	1986	3.40
1767	4.62	1811	−2.78	1855	2.94	1899	1.15	1943	3.33	1987	4.17
1768	−1.47	1812	13.57	1856	0.00	1900	4.55	1944	2.82	1988	4.90
1769	−7.46	1813	2.52	1857	−4.76	1901	0.00	1945	2.75	1989	7.78
1770	0.00	1814	−12.88	1858	−9.00	1902	0.00	1946	3.05	1990	9.46
1771	8.06	1815	−10.56	1859	−1.10	1903	1.09	1947	7.04	1991	5.87
1772	10.45	1816	−8.66	1860	3.33	1904	0.00	1948	7.61	1992	3.74
1773	0.00	1817	13.79	1861	2.15	1905	0.00	1949	2.89	1993	1.59
1774	1.35	1818	0.00	1862	−2.11	1906	0.00	1950	3.13	1994	2.41
1775	−6.67	1819	−2.27	1863	−3.23	1907	1.08	1951	9.09	1995	3.47
1776	−1.43	1820	−9.30	1864	−1.11	1908	0.00	1952	9.17	1996	2.41
1777	0.00	1821	−11.97	1865	1.12	1909	1.06	1953	3.05	1997	3.14
1778	2.90	1822	−13.59	1866	5.56	1910	1.05	1954	1.98	1998	3.43
1779	−8.45	1823	6.74	1867	6.32	1911	0.00	1955	4.36	1999	1.54
1780	−3.08	1824	8.42	1868	−0.99	1912	3.13	1956	5.10	2000	2.96
1781	4.76	1825	17.48	1869	−5.00	1913	−1.01	1957	3.53	2001	1.77
1782	1.52	1826	−5.79	1870	0.00	1914	0.00	1958	3.20	2002	1.67
1783	11.94	1827	−6.14	1871	1.05	1915	12.24	1959	0.41	2003	2.89
1784	1.33	1828	−2.80	1872	4.17	1916	18.18	1960	1.03	2004	2.98
1785	−5.26	1829	−0.96	1873	4.00	1917	25.38	1961	3.46	2005	2.84
1786	0.00	1830	−3.88	1874	−3.85	1918	22.09	1962	4.33	2006	3.18
1787	0.00	1831	10.10	1875	−2.00	1919	10.05	1963	1.89	2007	4.29
1788	4.17	1832	−7.34	1876	0.00	1920	15.53	1964	3.33	2008	4.00
1789	−1.33	1833	−5.94	1877	−1.02	1921	−8.70	1965	4.66	2009	−0.50
1790	1.35	1834	−8.42	1878	−2.06	1922	−13.85	1966	3.94	2010	4.60
1791	0.00	1835	2.30	1879	−4.21	1923	−6.03	1967	2.64	2011	5.20
1792	1.33	1836	11.24	1880	3.30	1924	−0.53	1968	4.65		
1793	2.63	1837	2.02	1881	−1.06	1925	0.00	1969	5.37		
1794	8.97	1838	0.99	1882	1.08	1926	−0.54	1970	6.40		

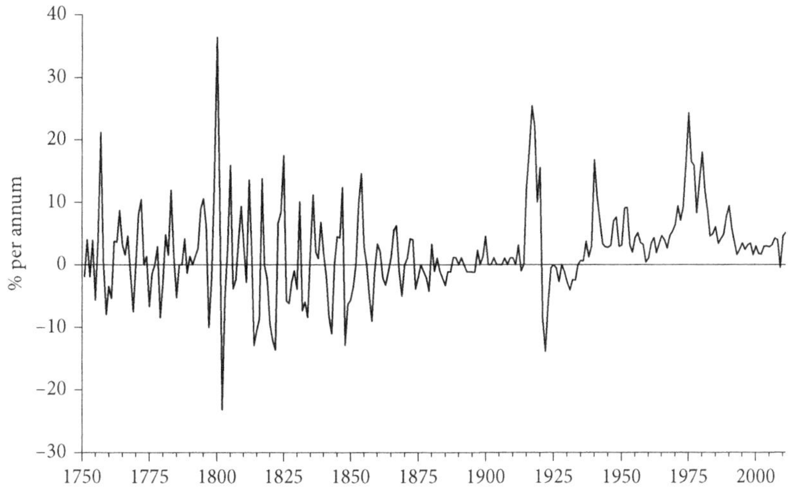

FIGURE 2.2 *Time series plot of UK inflation, 1751–2011*

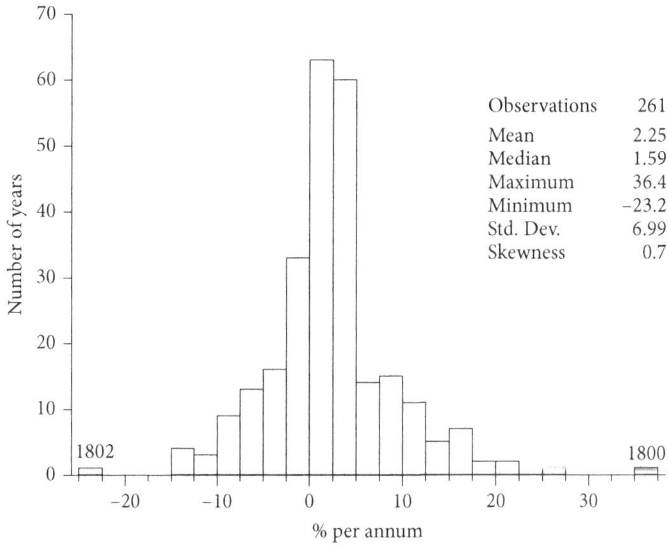

FIGURE 2.3 *Histogram of UK inflation*

$$\bar{x} = \frac{x_1 + x_2 + \ldots + x_N}{N} \tag{2.1}$$

Thus, if we have $N = 5$ observations, say $x_1 = 1$, $x_2 = 5$, $x_3 = 9$, $x_4 = 6$ and $x_5 = 2$, then the sample mean is

$$\bar{x} = \frac{1+5+9+6+2}{5} = \frac{23}{5} = 4.6 \tag{2.2}$$

For the data in Table 2.1, $N = 189$. With such a 'large' sample size, the notation used in equations (2.1) and (2.2) becomes unwieldy, so we replace it with *summation notation*, in which the sum $x_1 + x_2 + x_N$ is replaced by

$$\sum\nolimits_{i=1}^{N} x_i$$

which is to be read as 'the sum of the x_i from $i = 1$ to $i = N$': a typical data value being referred to as x_i. Where there is no confusion, the limits of the summation and the identifying subscript are often omitted, to give $\sum x$ (read as 'sigma x'). Thus

$$\bar{x} = \frac{\sum_{i=1}^{N} x_i}{N} = \frac{\sum x}{N}$$

As we see from Figure 2.1,

$$\bar{x} = \frac{\sum_{i=1}^{189} x_i}{189} = 13389$$

so that mean income is \$13389. Typically, *no* data value will exactly equal the mean: the closest to it here is St. Lucia, with an income of \$13,079.

While the sample mean is a very popular and simple way to calculate a measure of *location* or *central tendency*, it can be overly influenced by extreme values. Suppose that, in our $N = 5$ example, the third observation had been 59 rather than 9. The sample mean would then have been calculated as $(73/5) = 14.6$, which is over twice as large as the second highest value of 6: $\bar{x}$ has thus been 'dragged' towards the outlier and is no longer representative of the central tendency of the data.

Looking at the income histogram in Figure 2.1, we have already identified two outliers, so that the sample mean may be being unduly influenced by these values. An alternative measure of central tendency is the *median*. This is calculated in the following way. The data is first *sorted* in ascending order (as in Table 2.1) to give the sequence of observations

$$x_{[1]}, x_{[2]}, \ldots, x_{[N]} \qquad \text{where} \qquad x_{[1]} \leq x_{[2]} \leq \ldots \leq x_{[N]}$$

that is, the smallest observation is denoted $x_{[1]}$ (this is Zimbabwe with an income of just $143), the second smallest $x_{[2]}$ (the Democratic Republic of Congo, $231) and so on, with the largest value thus denoted $x_{[189]}$ (Qatar, which has already been identified as having an income of $159,144). If N is odd, then the median is given by

$$x_{med} = x_{\left[\frac{N}{2}+\frac{1}{2}\right]}$$

that is, it is the value for which as many observations lie below it in size as lie above it: it is the 'middle' value. Thus for our $N = 5$ example

$$x_{[1]} = x_1 = 1 \quad x_{[2]} = x_5 = 2 \quad x_{[3]} = x_2 = 5 \quad x_{[4]} = x_4 = 6 \quad x_{[5]} = x_3 = 9$$

and

$$x_{med} = x_{[3]} = 5$$

Note that even if $x_3 = 59$, the median would be unaffected because the *ordering* of the observations remains the same as before. Consequently, the median should be less affected by outliers.

If N is even, the median formula needs to be modified to

$$x_{med} = \frac{1}{2}\left(x_{\left[\frac{N}{2}\right]} + x_{\left[\frac{N}{2}+1\right]} \right)$$

that is, to the average of the two 'middle' values.

But for the data in Table 2.1, $N = 189$ is odd, so that

$$x_{med} = x_{[95]} = 7318$$

Thus the median income is $7318, which is the income of Montenegro, some way below the mean income of $13,389.

When we have time series data, we typically denote the values as $x_1, x_2, \ldots, x_T$, where T is the length of the series. With a typical value being denoted as x_t, the mean is defined accordingly as

$$\bar{x} = \frac{\sum_{t=1}^{T} x_t}{T}$$

and the median (for odd T) as

$$x_{med} = x_{\left[\frac{T}{2}+\frac{1}{2}\right]}$$

Note that the median calculation does *not* preserve the natural time ordering of the data. Given the above notation, it is easy to see why cross-sectional data is sometimes referred to as 'little-i' data, and time series as 'little-t' data.

For the inflation data shown in Table 2.2, the mean is 2.25% and the median is 1.59%, so that they are much closer together than in the income data. A reason for this will be discussed shortly.

2.3 Summary statistics: measures of dispersion (variation)

While having information about the central tendency of a data set is generally useful, we usually also want a measure of how the observations are dispersed around that central value. Given our sorted data $x_{[1]}, x_{[2]}, \ldots, x_{[N]}$ (we use little-i notation for convenience), a simple measure of dispersion is the *range*, which is defined as the difference between the maximum and minimum values of the data set. Since the minimum is $x_{[1]}$ and the maximum is $x_{[N]}$, we have

$$range = x_{[N]} - x_{[1]}$$

Although clearly a very simple figure (it can readily be calculated from the information provided in Figures 2.1 and 2.3, for example), it suffers from being completely reliant on just the two extreme values of the data. An improvement is the *inter-quartile range*. The *quartiles* are those values that 'split' the ordered data into four equally sized parts. For N odd, the first and third quartiles are

$$Q_1 = x_{\left[\frac{N}{4}+\frac{3}{4}\right]} \qquad Q_3 = x_{\left[\frac{3N}{4}+\frac{1}{4}\right]}$$

For example, with $N = 189$, $Q_1 = x_{[48]}$ and $Q_3 = x_{[142]}$, it should be easy to see that the second quartile, Q_2, is, in fact, the median, here $x_{[95]}$. The inter-quartile range is then defined as

$$IQR = Q_3 - Q_1$$

The *IQR* thus defines the limits of the middle half (50%) of the distribution. If we have an even number of observations, the formulae for the quartiles need to be modified in an analogous way to that of the median. For our income inequality data, the quartiles are $Q_1 = x_{[48]} = \$2300$ (Kyrgyzstan) and $Q_3 = x_{[142]} = \$19233$ (Libya). Thus

$$IQR = 19233 - 2300 = 16933$$

On its own, the *IQR* is no more than a summary measure, but it could be used to compare the dispersion of, say, two income distributions if they were both measured in the same currency units: the distribution with the larger *IQR* would exhibit the greater inequality.

A more useful measure of dispersion is the *sample variance*, which makes use of all the available data. It is defined as the average of the 'squared deviations about the mean',

$$s^2 = \frac{\sum_{i=1}^{N}\left(x_i - \bar{x}\right)^2}{N-1}$$

Note that the divisor in s^2 is N–1 rather than N, as it would be for a 'true' average. The reason for this is quite technical, and will be discussed in **§11.1**. One difficulty with the variance is that it is measured in units that are the *square* of the units that the data are measured in. Thus, for our income data the variance will be in 'squared dollars', which are both very difficult to interpret and, in this case, lead to a very large numerical value for the variance. To get back to the original units we may take the square root of the variance, thus defining the *sample standard deviation* (denoted 'Std. Dev.' in Figures 2.1 and 2.3),

$$s = \sqrt{\frac{\sum_{i=1}^{N}\left(x_i - \bar{x}\right)^2}{N-1}}$$

so that for the income data $s = 17629$, while for the inflation data $s = 6.99$. Taken on their own, standard deviations are also of limited use. However, when related to the sample mean, they become a very important statistic for measuring a variety of features of the data. These ideas will be developed in due course, but a simple way of bringing together the sample mean and standard deviation is to define the *coefficient of variation* as the ratio of s to $\bar{x}$: $CV = s/\bar{x}$.

For the income data, $CV = 17629/13389 = 32$, whereas for the inflation data, $CV = 6.99/2.25 = 3.11$. Thus for the income data the standard deviation is 1.3 times the mean, but for the inflation data the ratio is over 3: in relative terms inflation in the UK displays much more variability than world income (with some fairly obvious qualifiers!).

2.4 Summary statistics: the Boxplot

We have introduced a variety of concepts and measures for summarising data sets. We now present a neat pictorial representation that incorporates many of these measures in one display: the *boxplot*. Figure 2.4 shows a boxplot for the income inequality data.

Income is measured on the vertical axis. The rectangular box stretches (vertically) from the first to the third quartile and thus has a height equal to the inter-quartile range, covering the central half of the distribution. The mean and median are indicated by a dot and a horizontal line respectively. Two 'whiskers' extend above and below the box as far as the highest and lowest observations *excluding outliers*. These are defined as any observation more than 1.5 times the inter-quartile range above or below the box. Observations lying between $1.5 \times IQR$ and $3 \times IQR$ above or below the box are termed *near outliers*; any more than $3 \times IQR$ are *far outliers*.

Luxembourg and Qatar are shown to be far outliers, while a group of seven countries (numbers 181–187 in Table 2.1) are identified as near outliers. There are no outliers below the box, so that the end of the lower whisker represents the minimum income value, that of Zimbabwe.

The corresponding boxplot for UK inflation is shown in Figure 2.5.

We see that there are many more outliers than for the income data set: indeed, so many that we only identify the maximum and minimum and do not distinguish between far and near outliers. This boxplot confirms that the variability of the inflation data is much greater than that of the income data.

2.5 Summary statistics: symmetry and skewness

An important data feature that we have yet to comment upon is that of the *symmetry*, or indeed the *asymmetry*, of the underlying frequency distribution.[4] Consider again the histograms shown in Figures 2.1 and 2.3. Symmetry is the property of the 'left-hand side' of a frequency distribution being the mirror image of the 'right-hand side'. Of course, this raises the question of how we define left- and right-hand sides of a distribution. We may do this using a further measure of location, the *mode*.

With 'raw' data, like those tabulated in Tables 2.1 and 2.2, the mode is a particularly useless measure, as it is defined to be the value that occurs most often. For both sets of data, no value occurs more than once (particularly if we were to report the inflation data to several decimal places), so that there are N different values for the mode. Where the measure does become useful is when the data are *grouped* into a histogram. We can then see from Figures 2.1 and 2.3 that the *modal class* is sensibly defined to be the class that occurs most frequently: $0–2500 for income and 0–2.5 per cent for inflation. A distribution is then said to be symmetric if, centred on the modal class, the left-hand side of the histogram is the mirror image of the right-hand side.

Of course, it is unlikely that a histogram will be perfectly symmetric, so we tend to be interested in how asymmetric a distribution is: that is,

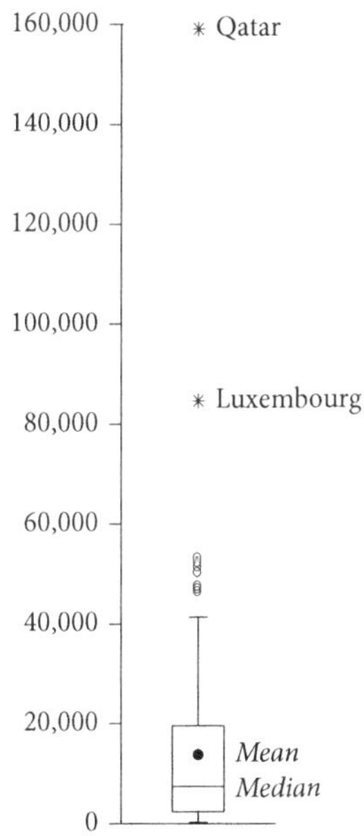

FIGURE 2.4 *Boxplot for real per capita GDP, 2009*

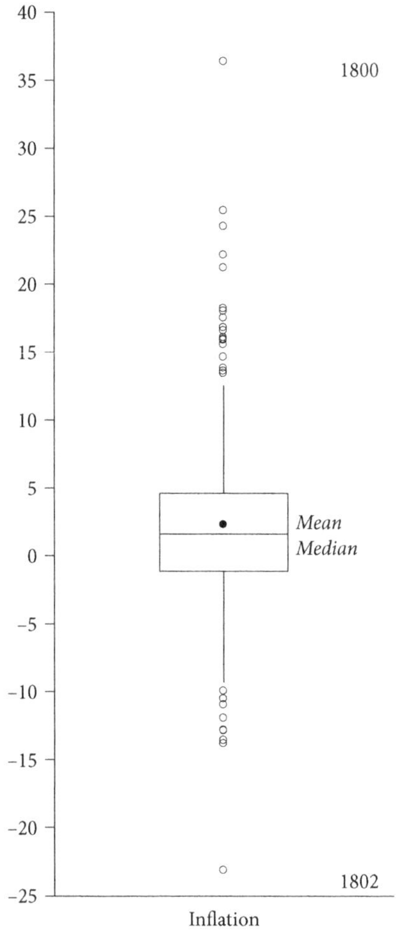

FIGURE 2.5 *Boxplot for UK inflation, 1750–2011*

we would like a measure of the extent of asymmetry, known as *skewness*. One such measure is the *coefficient of skewness*, defined as

$$skew = \frac{\sum_{i=1}^{N}(x_i - \bar{x})^3}{Ns^2}$$

This may be interpreted as the average of the '*cubed* deviations about the mean' divided by the sample variance, which ensures that it is a 'scale free' measure. If the large positive deviations about the mean outweigh the large negative deviations, then *skew* will be positive, and the distribution is said to be skewed to the right. In such a case, the sample mean will be larger than the median, which in turn will be larger than the mode. If the converse holds then the distribution will be negatively skewed. A value of zero will signify a perfectly symmetric distribution.

From Figure 2.1 we see that *skew* = 3.8, so that the income distribution is positively skewed. This obviously follows from the fact that the modal class is also the smallest, reflecting both that income cannot be negative and that many countries have very low incomes compared to the relatively fewer extremely wealthy nations. From Figure 2.3, inflation has *skew* = 0.7, so again the distribution is positively skewed. However, unlike income, inflation can go negative, but the positive deviations from the mean nevertheless still outweigh the negative deviations, reflecting the tendency for inflation to be positive, so that prices generally rise over time (for more detail on the relationship between prices and inflation, see §3.2). It is easily seen that the inequality mean > median > mode holds in both examples, but that the smaller skewness in the inflation data leads to the three measures being (relatively) closer together than for the income data. The boxplots reflect positive skewness by having more outliers above the box than below it, and showing that the mean exceeds the median.

2.6 Scatterplots

So far, we have been considering data observed on one variable in isolation from data on other, possibly related, variables. Economics typically constructs theories relating one variable to another or, more usually, to a set of variables. When two variables are thought to be related, it is usually informative to construct a *scatterplot*. Suppose we have data on two variables, Y and X, and a particular economic theory suggests that movements in X produce movements in Y: in other words, the *dependent variable* Y is some function, call it $f(\ )$, of the *independent variable* X, so that $Y = f(X)$. The data on the two variables can now be regarded as coming in (ordered) pairs:

$$(y_1, x_1),\quad (y_2, x_2),\quad \ldots,\quad (y_N, x_N)$$

These pairs may be plotted as a scatterplot with the convention that the dependent variable is plotted using the vertical axis and the independent variable the horizontal axis.

Interest rates and inflation in the UK

The main instrument of current UK macroeconomic policy is the setting of interest rates by the Bank of England's Monetary Policy Committee (MPC). The interest rate is set with the primary aim of meeting a 2% inflation target 18 months ahead. An interesting question is to examine the historical link between inflation and interest rates. Figure 2.6 presents a scatterplot of the long interest rate (the yield on gilt edged stock) and inflation for the period 1751 to 2011.

An intriguing picture emerges: except for the period from 1965 to 1997 (shown as dots), at the start of which the Bank of England was granted independence and the current inflation-targeting arrangements came into practice, there looks to be only a very weak link between the two variables, with low (less than ~6%) interest rates being associated with a wide range of inflation rates. During the period from the mid-1960s to the mid-1990s, however, there is a clear positive link between interest rates and inflation: this was the era in which inflation was high, persistent and volatile, and interest rates reflected that. But for the bulk of the

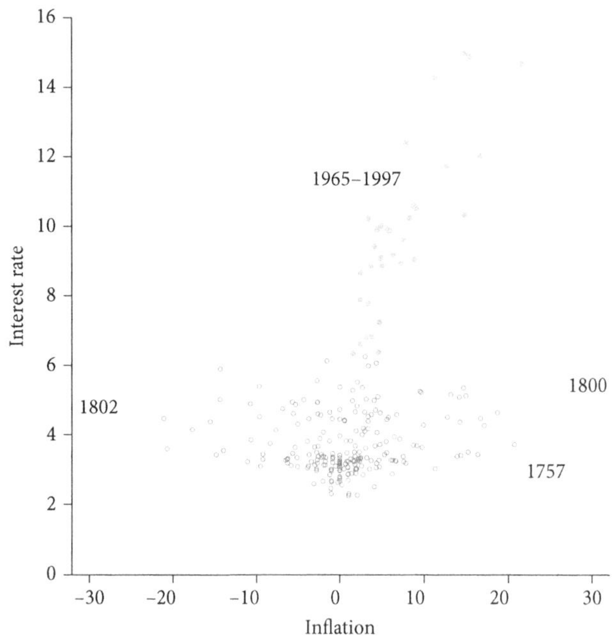

FIGURE 2.6 *Scatterplot of UK long interest rate and inflation, 1751–2011*

two and a half centuries, no strong link is observed. This could be due to a variety of reasons. Obviously, there could simply be no relationship and interest rates were set according to other factors, such as the need to increase government borrowing to fund expenditure on wars. A more subtle reason probably holds since 1997: the MPC have been so successful in targeting inflation that they have kept the rate at ~2% by shifting interest rates around, so that there does not appear to be any link between the two variables!

This argument, though, is looking increasingly implausible since the global financial turmoil of 2008. Since then, interest rates have been kept artificially low, while inflation has remained well above target, arguably partly a consequence of the Bank of England's policy of 'quantitative easing', as well as external factors such as rapid increases in energy prices and a fall in the sterling exchange rate. The relationship between these two variables is analysed in considerably more detail and formality in §17.5.

This example nevertheless illustrates, among other things, three key features of what we might call *exploratory data analysis*:

i) the importance of relationships shifting within the sample, possibly due to institutional arrangements changing;
ii) how subtle and imprecise relationships between economic variables can be; and
iii) how tricky it can be to decide which variable is the dependent and which is the independent in any particular analysis – do changes in inflation always produce changes in interest rates, or is it the other way round or, indeed, is it even a mixture of both?

2.7 Extensions

The various measures of location, dispersion and skewness all have counterparts that can be used when the data come in the form of grouped observations – if, say, all we have available are the class widths and the numbers of observations falling in each class (in other words, a tabular representation of the histograms in Figures 2.1 and 2.3).

The form of boxplot presented here is the simplest possible, and various modifications can be made to convey more sophisticated aspects of the data set. Scatterplots are, of course, restricted to depicting the relationship between just two variables, but we are often interested in the links between several variables. There are many ways of extending scatterplots to cope with data in several dimensions, although of necessity these tend to require specialised software and will not be discussed further here.

Notes

1 The standard textbook reference to panel data is Badi H. Baltagi, *Econometric Analysis of Panel Data*, 4th edition (Chichester, Wiley, 2008).
2 The income data can be found in the Penn World Tables, located at http://pwt.econ.upenn.edu/php_site/pwt_index.php (Alan Heston, Robert Summers and Bettina Aten, *Penn World Table Version 7.0*, Center for International Comparisons of Production, Income and Prices at the University of Pennsylvania, May 2011).
3 The original source for the inflation data is Jim O'Donoghue, Louise Goulding and Grahame Allen, 'Consumer price inflation', *Economic Trends* 604 (2004), 389–413.
4 Symmetry has a much wider importance than in data analysis alone, being at the basis of communication and evolutionary biology. For a fascinating account of the wider nature of symmetry, see Marcus du Sautoy, *Finding Moonshine: A Mathematician's Journey through Symmetry* (London, Fourth Estate, 2008).

3
Transforming Data

Abstract: *The need to often transform raw data is discussed and the logarithmic transformation is introduced in some detail. It is emphasised that the slope of a graph of the original data says nothing about the growth rate of the variable, since it is only from the slope of the graph of the logarithms that such information can be obtained. These ideas are illustrated by constructing alternative measures of UK inflation. Other transformations are discussed, including the famous Phillips curve, linking inflation to the inverse of the unemployment rate. Moving averages are introduced as a way of smoothing data and such ideas are extended to decomposing a time series, illustrated by decomposing retail sales into its trend, seasonal and irregular components as a prelude to seasonally adjusting the series.*

3.1 Why transform data?

It is often the case that we do not analyse the 'raw' data on a particular variable or set of variables, but mathematically manipulate the numbers (in general, *transform* the data) into a form that we consider to be more amenable for analysis. Why should this be? One reason is that, occasionally, economic theory suggests the mathematical form that the variables should take. For example, the *Cobb–Douglas production function* links output, Y, to capital and labour inputs, K and L, respectively, by the relationship

$$Y = AK^{\alpha}L^{\beta} \tag{3.1}$$

Multiplicative relationships like (3.1) can be tricky to handle, so we *linearise* (straighten out) the function by taking logarithms to yield[1]

$$\ln Y = \ln A + \alpha \ln K + \beta \ln L \tag{3.2}$$

The production function is now linear in the transformed variables ln Y, ln K and ln L, and is much easier to handle both mathematically and statistically. As we shall see, other reasons for transforming data are essentially statistical and are often suggested during the exploratory stage of data analysis.

3.2 The logarithmic transformation

The logarithmic transformation used above is employed regularly in economics. One important use is to linearise time series that are growing at constant rates. To illustrate this, consider the plots in Figures 3.1(a) and (b). Figure 3.1(a) shows a series, Y_t, that is growing at a constant rate of 10% per period.

It is calculated using the equation

$$Y_t = 1.10Y_{t-1} \tag{3.3}$$

which is the mathematical formula for generating a sequence that increases by 10% each period: the start-off point was chosen to be $Y_1 = 10$, so that

$$Y_2 = 1.10 \times Y_1 = 11,$$

$$Y_3 = 1.10 \times Y_2 = (1.10)^2 Y_1 = 12.1,$$

$$Y_4 = 1.10 \times Y_3 = (1.10)^3 Y_1 = 13.31, \text{etc.}$$

Thus, by $t = 30$, say, $Y_{30} = (1.10)^{29}\ Y_1 = 158.63$. Note that, although the growth rate is *constant*, the slope of the function is *increasing* over time: it is thus *incorrect* to interpret the slope of a plot of the *levels* of Y_t against time as a *growth rate*.

Figure 3.1(b) shows the logarithms of the series:

$$\ln Y_1 = \ln 10 = 2.303,$$

$$\ln Y_2 = \ln 11 = 2.398,$$

$$\ln Y_3 = \ln 12.1 = 2.493,$$

$$\ln Y_4 = \ln 13.31 = 2.588, \text{etc.}$$

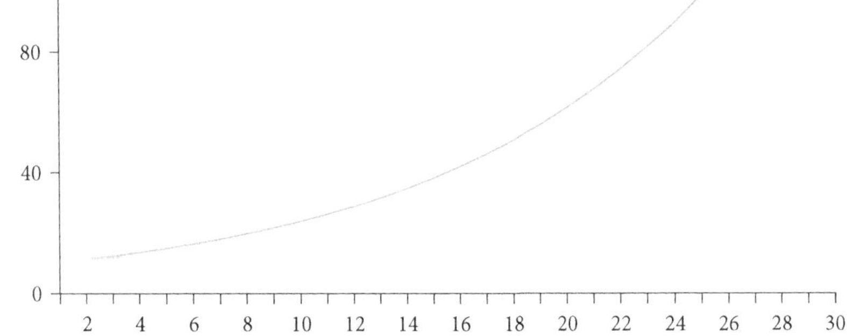

FIGURE 3.1(a) *Example of a time series growing at a constant rate of 10% generated as* $Y_t = 1.10Y_{t-1}$, $Y_1 = 10$

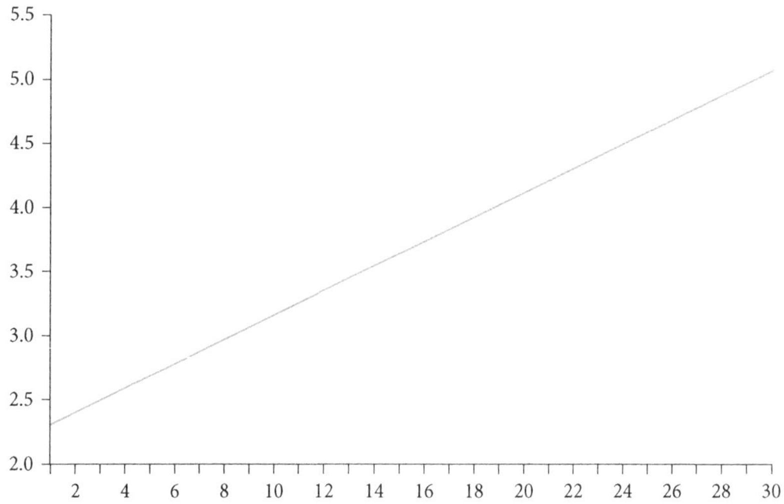

FIGURE 3.1(b) *Logarithm of* Y_t: $\ln Y_t$

The slope is now constant at $\ln 1.10 = 0.0953$, showing that you have to plot the logarithms of the series before you can infer anything about growth rates from the slope. Note also that $\ln Y_{30} = 5.067$, so that the range of the observations has been reduced from ~10–160 to ~2.3–5: taking logarithms thus *compresses the scale* of the data, and this can make analysis and interpretation simpler.

Now, suppose that a series takes the values 10, 11, 12 and 13 at times $t = 1, 2, 3$ and 4, that is, it changes by 1 every period and hence is a linear function of time: it should be easy to see that $Y_t = 9 + t$.

What are its growth rates, though? The standard formula to calculate the growth rate (in percentages) of the series Y_t between periods t–1 and t is

$$g_t = 100\frac{Y_t - Y_{t-1}}{Y_{t-1}} \tag{3.4}$$

This formula may be derived from the general form of (3.3) in the following way. For a growth rate g_t, (3.3) becomes

$$Y_t = \left(1+\frac{g_t}{100}\right)Y_{t-1} \tag{3.5}$$

which of course gives (3.3) exactly if $g_t = 10$.

Equation (3.5) may be written as

$$\frac{Y_t}{Y_{t-1}} = 1+\frac{g_t}{100}$$

or as

$$g_t = 100\left(\frac{Y_t}{Y_{t-1}}-1\right) = 100\frac{Y_t - Y_{t-1}}{Y_{t-1}}$$

which is indeed (3.4). Thus, the growth rate between $t = 1$ and $t = 2$ is

$$g_2 = 100\frac{11-10}{10} = 10\%,$$

but the growth rate between $t = 2$ and $t = 3$ is only

$$g_3 = 100\frac{12-11}{11} = 9.09\%$$

even though the *change* remains the same at 1 unit. Similarly, the next growth rate, g_4, is 8.33%. A time series that changes by a constant amount each period therefore exhibits *declining* growth rates over time, because the denominator in (3.4) is increasing each period, whereas the numerator remains constant.

Note also that the constant slope of the logarithmic function implies that, by taking logarithms of (3.3), $\ln Y_t - \ln Y_{t-1} = \ln 1.10$. On taking logarithms of (3.5) we have, if we assume for simplicity that $g_t = g$, a constant,

$$\ln Y_t - \ln Y_{t-1} = \ln\left(1+\frac{g}{100}\right) = \ln(1+x) \tag{3.6}$$

where $x = g/100$ must be a positive but typically small number (if $g = 10$, $x = 0.1$). The logarithmic series expansion under these conditions is

$$\ln(1+x) = x - \frac{x^2}{2} + \frac{x^3}{3} - \frac{x^4}{4} + \ldots$$

Now, for small x, the terms in the expansion containing x^2, x^3, … etc. will all be much smaller than x, so that the expansion can be approximated as

$$\ln(1+x) \approx x \tag{3.7}$$

Thus, since x is the growth rate measured in decimals, equating (3.7) with (3.6) gives

$$\ln Y_t - \ln Y_{t-1} \approx x$$

or

$$g = 100x \approx 100(\ln Y_t - \ln Y_{t-1}) \tag{3.8}$$

that is, the change in the logarithms (multiplied by 100) is an estimate of the percentage growth rate at time t. Such a change is often denoted by an upper-case delta, Δ, for example, $\ln Y_t - \ln Y_{t-1} = \Delta \ln Y_t$.

We thus have the approximate equivalence (now making it explicit that the growth rate can change over time by using g_t)

$$g_t = 100\frac{Y_t - Y_{t-1}}{Y_{t-1}} = 100\left(\frac{Y_t}{Y_{t-1}} - 1\right) \approx 100(\ln Y_t - \ln Y_{t-1}) = 100 \ln\left(\frac{Y_t}{Y_{t-1}}\right)$$

as long as $(Y_t/Y_{t-1}) - 1$ is small. For a growth rate of 10%, using the change in the logarithms to approximate this rate gives $100\ln(1.10) = 9.53\%$, which may or may not be sufficiently accurate for the purpose at hand. For a smaller growth rate, say 2%, $100\ln(1.02) = 1.98\%$ gives a much more accurate approximation, as predicted (with $x = 0.1$, the ignored second term in the logarithmic series expansion is 0.005 or 0.5%, but with $x = 0.02$ it is only 0.0002, or 0.02%).

Thus taking logarithms not only linearises a growing time series, but the successive changes in the logarithms can under many circumstances be used as estimates of the growth rate of the series over time.

Calculating UK inflation

One of the most important growth rate computations in economics is the calculation of the rate of inflation. Figure 3.2 plots the monthly UK retail price index from January 1948 to December 2011.

This plot is typical of an economic time series that is generally growing at a positive, but not constant, rate.

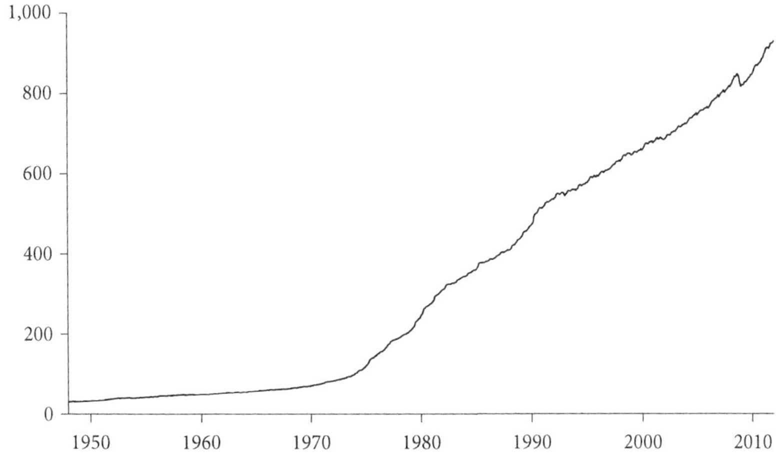

FIGURE 3.2 *UK retail price index, P_t; monthly, 1948–2011*

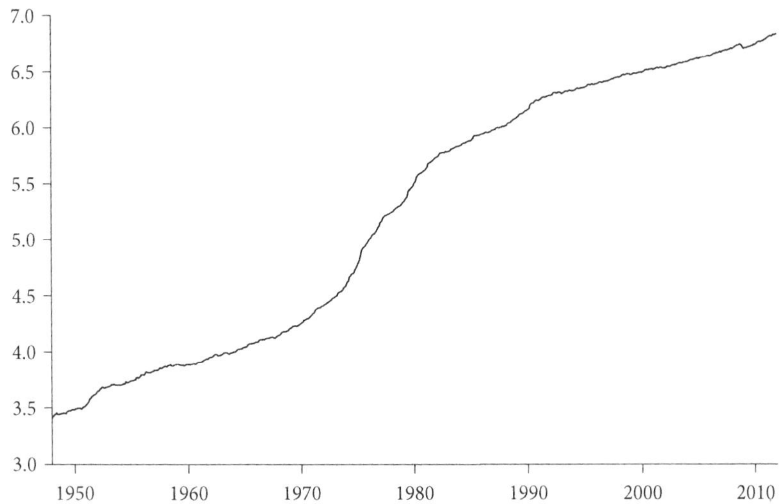

FIGURE 3.3 *Logarithms of the UK retail price index, $\ln P_t$; monthly, 1948–2011*

The plot is certainly not linear, so in Figure 3.3 we show the logarithms of the series. The slopes of the pre-1970 and post-1990 observations are fairly similar, but the intervening two decades are characterised by a much steeper slope.

We emphasise at this point that *neither* of the series shown in the two figures can be interpreted as the rate of inflation. Figure 3.2 is a plot of the *price level*, which we denote as P_t, while Figure 3.3 is a plot of the *logarithms of the price level*, $\ln P_t$.

How can we calculate the rate of inflation? From the discussion above, an obvious way would be to use the growth rate formula (3.4), or its approximation (3.8), which here would be

$$\pi_t^m = 100\frac{P_t - P_{t-1}}{P_{t-1}} \approx 100\left(\ln P_t - \ln P_{t-1}\right) \tag{3.9}$$

The notation π_t^m is used to signify that we are calculating the *monthly rate of inflation*, that is, computing the rate at which prices change from one month to the next. Figure 3.4 plots this rate of inflation scaled up by a factor of 12 (for reasons that will become clear shortly). It is an extremely volatile series, ranging from a minimum of −20% to a maximum of 52%. No discussions of contemporary UK macroeconomics (or, indeed, of any era of the UK economy) have mentioned such huge variations in inflation which, if they had occurred, would surely have led to unprecedented economic and social dislocation. Yet we have used the standard formula for calculating a growth rate. So what has, on the face of it, gone wrong?

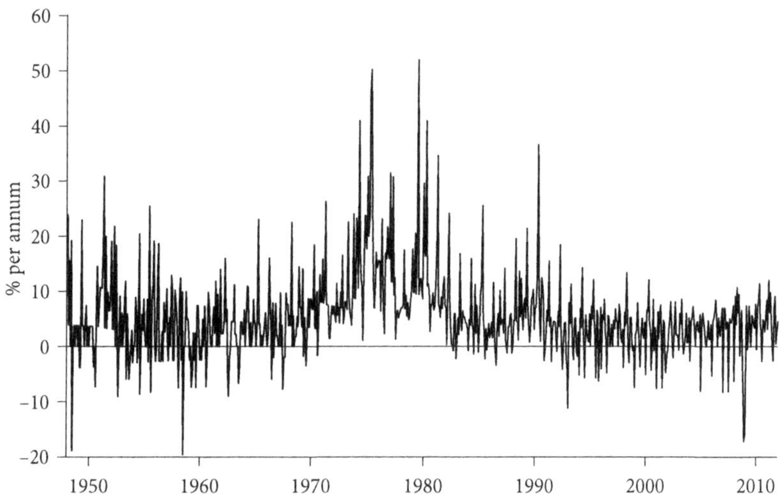

FIGURE 3.4 *Monthly annualised UK inflation, π_t^m, 1948–2011*

One reason for the very large (absolute) inflation values is the scaling up by a factor of 12: this has been done to provide an *annualised monthly rate of inflation*. Could we just ignore this scaling factor? Yes, of course we could, but the variation from one month to the next would still be substantial and this is not what contemporary discussions of macroeconomic performance focus upon.

We can avoid the problems found in Figure 3.4 by *calculating annual rates of inflation*

$$\pi_t^a = 100\frac{P_t - P_{t-12}}{P_{t-12}} \approx 100\left(\ln P_t - \ln P_{t-12}\right) \tag{3.10}$$

where the rate of inflation is calculated by comparing the price level at time t with the level that occurred one year (12 months) previously; for example the December 2011 price level is compared with the December 2010 level. The time series for π_t^a is shown in Figure 3.5, and is much smoother and less volatile than that for π_t^m. The two rates are on the same scale, though, so that comparisons are valid, which was why π_t^m needed to be scaled by a factor of 12.

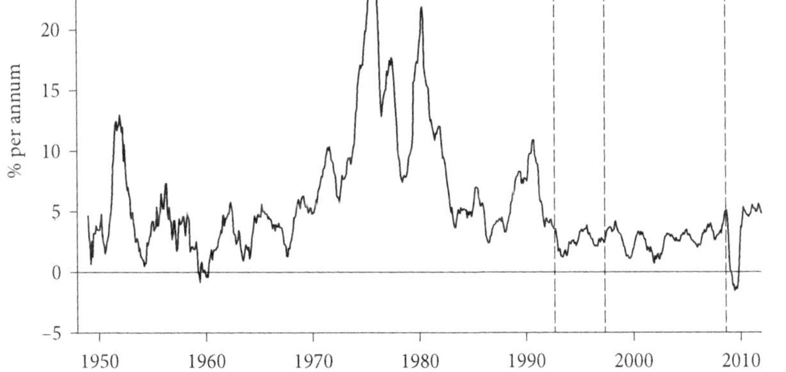

FIGURE 3.5 *Annual UK inflation, π_t^a, 1949–2011*

There is thus *no unique definition* of inflation, as it can be calculated in various ways (try defining the *annualised quarterly rate of inflation*). Thus when defining the rate of inflation we should be very careful that we actually calculate the rate that we wish to use. The definition (3.10) is the one used by the MPC, the Office of National Statistics and by commentators of the UK economy, while (3.9) has traditionally been favoured in the US.

Figure 3.5 shows that UK inflation has gone through several distinct phases since the end of World War II. Apart from a short period in the early 1950s (an era of 'cheap money' and low interest rates, designed to increase growth and living standards after the war), inflation was generally low and stable up till the mid-1960s. For the next two decades inflation was high and volatile, a consequence of two major external oil price shocks and an often lax internal monetary policy. During the mid-1980s inflation was again quite low, but this was only temporary and it took off again at the end of that decade. Since the exit from the European Exchange Rate Mechanism (ERM) in October 1992, inflation has been low and relatively stable, at least compared with the previous two decades: Bank of England independence from 1997 has merely continued the inflation 'regime' that had been in operation since the Black (or should it really be White?) Wednesday of the ERM exit. Note the decline in inflation during the credit crunch of 2007–2008, which invoked for a short while fears of *deflation* – a sustained period of falling prices – although since 2009 inflation has returned and has remained stubbornly above the Bank of England's target range.

3.3 Other transformations

Figure 3.6 shows the shapes of three functional forms commonly found to link pairs of variables in economics.

The first two use an *inverse* transformation on the independent variable X (that is, $1/X$), while the third just transforms Y logarithmically (the *semi-log* functional form). Various other functional forms based on the transformation of variables may easily be conceived of.

The Phillips curve

One of the most famous examples of applied economics was published in the journal *Economica* by A.W. (Bill) Phillips in 1958.[2] This investigated

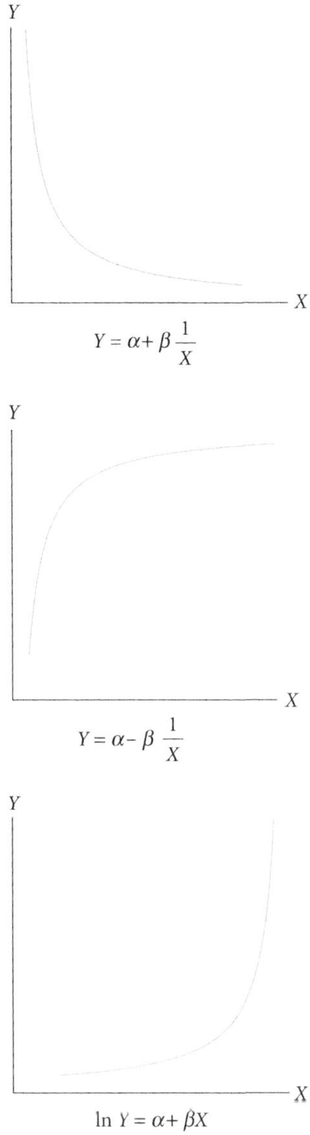

FIGURE 3.6 *Alternative functional forms*

the relationship between the rate of inflation (π) and the unemployment rate (U) in the UK from 1861 to 1957, and found that a curve of the form

$$\pi = \alpha + \beta \frac{1}{U} \qquad \alpha > 0,\ \beta > 0 \tag{3.11}$$

explained the relationship well. This is an example of the first transformation in Figure 3.6 and implies that there is a trade-off between inflation and unemployment: to get inflation low, unemployment has to be high and vice versa. The implications for economic policy of such a trade-off have dominated macroeconomics ever since (although we will not discuss this here!).

Figure 3.7 updates the original Phillips data set to 2011. It superimposes a fitted line of the form (3.11): how such a line is fitted will be discussed later, in Chapter 6, but it is calculated to be

$$\pi = 1.4 + 4.4\frac{1}{U} \tag{3.12}$$

that is, $\alpha = 1.4$ and $\beta = 4.4$. To get a feel for how to interpret this equation, inflation and unemployment since 2000 have averaged 3% and 5.9% respectively. Equation (3.12) implies that, if $U = 5.9$, then inflation should be

$$\pi = 1.4 + \frac{4.4}{5.9} = 2.1$$

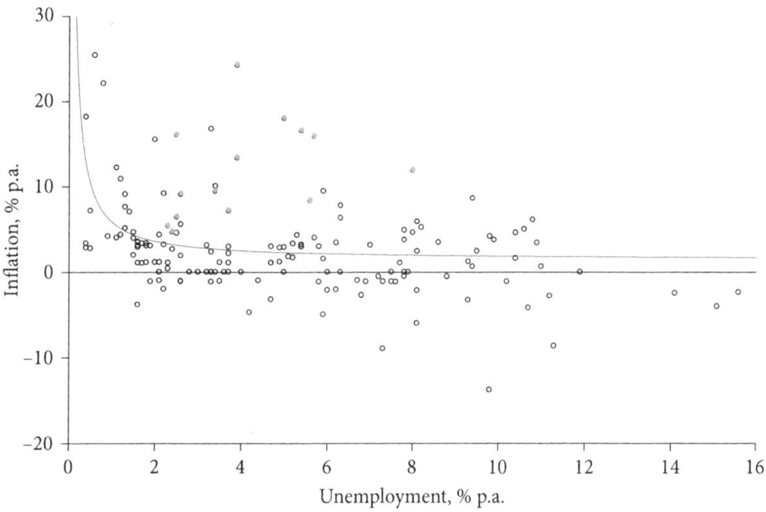

FIGURE 3.7 *Phillips curve fitted to UK inflation and unemployment, 1855–2011, both measured in % p.a. Dots, rather than open circles, signify observations between 1968 and 1981*

so that during the first decade of the 21st century inflation has been somewhat higher, on average, than predicted by the Phillips curve. However, if we wanted to get inflation down to 2%, the Bank of England's target, then unemployment would need to rise to 7.3%, obtained by solving the equation

$$2 = 1.4 + \frac{4.4}{U}$$

The very flat part of the fitted Phillips curve implies that further declines in inflation would require higher and higher rates of unemployment: 1.5% inflation would need 44% unemployment, higher than any level reached in the past and certainly unacceptable to any government (and electorate!).

However, we should treat this fitted curve with some scepticism. The fit is quite loose, with many observations lying some distance from the line: we shall propose a measure of 'goodness of fit' in §6.4. Furthermore, for some historical periods the Phillips curve does not seem to hold at all. The observations shown with dots are those between 1968 and 1981: a curve fitted to just these observations is $\pi = 19.2 - 25.9/U$, which has a *positive* slope (see Figure 3.6), implying that high rates of unemployment are accompanied by high, rather than low, inflation. This is a period that is often referred to as an era of *stagflation*, a term defined as the concatenation of *stagnation* (high unemployment) and *inflation*.

3.4 Moving averages

With time series data, we often want to focus on long-term (permanent) movements without the eye being distracted by short-run (transitory) fluctuations. We thus often want to *smooth* the data, and while there are many very sophisticated ways of doing this the easiest method is to calculate a *moving average*. The simplest of these is the three-period, equal weighted, centred moving average, defined for a time series x_t as

$$MA_t(3) = \frac{x_{t-1} + x_t + x_{t+1}}{3}$$

that is, each x_t is replaced by the average of itself, its previous value, x_{t-1}, and its next future value, x_{t+1}. The more future and past values that are included in the moving average, the smoother will be the resulting series. In general, we can define the $2n+1$-period moving average as

$$MA_t(2n+1) = \frac{x_{t-n} + \ldots + x_{t-1} + x_t + x_{t+1} + \ldots + x_{t+n}}{2n+1} = \frac{\sum_{i=-n}^{n} x_{t+i}}{2n+1} \tag{3.13}$$

Clearly n observations will be 'lost' at the start and end of the sample, and each of the values in the moving average can be thought of as having the *weight* $1/2n+1$ attached to it. The set-up of (3.13) ensures that the MA is computed over an odd number of observations, and its symmetric nature enables $MA_t(2n+1)$to 'match up' with (to be centred on) x_t. If an even number of terms is used in the moving average, then this centring will not happen unless an adjustment is made. For example, an MA of order four may be defined as

$$MA_{t-1/2}(4) = \frac{x_{t-2} + x_{t-1} + x_t + x_{t+1}}{4}$$

The notation used makes it clear that the 'central' date to which the moving average relates to is non-integer, being halfway between $t-1$ and t, that is $t-1/2$ – but of course $x_{t-1/2}$ does not exist! At $t+1$, however, this moving average is

$$MA_{t+1/2}(4) = \frac{x_{t-1} + x_t + x_{t+1} + x_{t+2}}{4}$$

which has a central date of $t+1/2$.

Taking the average of these two moving averages produces (3.14) below, centred on the average of $t-1/2$ and $t+1/2$, which is, of course, t.

$$WMA_t(5) = \frac{1}{8}x_{t-2} + \frac{1}{4}x_{t-1} + \frac{1}{4}x_t + \frac{1}{4}x_{t+1} + \frac{1}{8}x_{t+2} \tag{3.14}$$

When compared to (3.13) with $n = 2$, and hence equal weights of $1/5$, (3.14) is seen to be a *weighted* moving average (WMA) with 'half-weights' on the two extreme observations.

In general, a weighted moving average can be defined as

$$WMA_t(2n+1) = \sum_{i=-n}^{n} \omega_i x_{t+i} \qquad \sum_{i=-n}^{n} \omega_i = 1$$

Many *trend extraction techniques* lead to WMAs of various types, but these are too advanced to be discussed in any detail here.[3]

$-£ exchange rate

The $-£ exchange rate is, along with the €-£ exchange rate, the most important currency in the London Foreign Exchange (FOREX) market, and its movements are followed closely by traders and commentators alike. Figure 3.8 plots the $-£ rate from January 1973, just as it began to float freely after the breakdown of the Bretton Woods system of international finance, until December 2011.

Exchange rates typically follow quite variable and volatile time paths (they are close to being what are known as *random walks*: see §6.6) and it is often useful to smooth out short-run fluctuations to be able to concentrate on longer-run movements, known as *long swings*. To do this, we calculate an $MA(13)$ and superimpose it on the plot. An order of 13 was chosen because it will smooth out fluctuations that last for a year or less. The long swings in the exchange rate can now be seen somewhat more easily, with the rate appearing to go though extended periods of appreciation and depreciation.

A close look at the figure reveals that $n = 6$ values of the MA are lost at the start and end of the sample period. While this is usually not important at the start of the sample, as these observations are typically a long time in the past, the loss of values indicating the current trend of a time series at the end of the sample can be a major disadvantage. There are ways

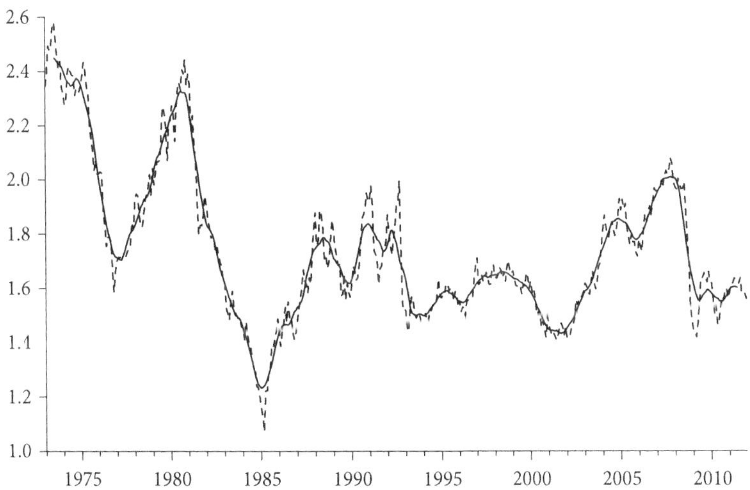

FIGURE 3.8 *$-£ exchange rate; January 1973 to December 2011, with 13-month moving average trend*

of overcoming this problem, typically by using some form of weighted moving average whose weights adjust as we get closer to the end of the sample – but again, this is too advanced a topic to be discussed here. The next example illustrates, without going into technical detail, one possible solution to the problem.

Global temperatures

This example concerns a series that, while not being an economic variable itself, may, if current concerns over global warming and climate change are substantiated, exert a major influence over the economic performance of many countries in years to come.

Figure 3.9 shows the annual global temperature series from 1850 to 2011. Superimposed upon it is an estimated trend, known as the *Hodrick–Prescott (trend) filter*, which is essentially a weighted moving average with end-point adjustments that allow the trend to be calculated right up to the end of the sample.[4] This, of course, is important here, because global temperatures are quite volatile, and extracting the recent trend is thus essential for providing an indication of the current extent of global warming. We see that the series was on a rising trend from around 1970 to 1998, but has since levelled off. Before 1970, the trend went through long swings, with cooling trends between 1875 and 1910 and between 1945 and 1970.

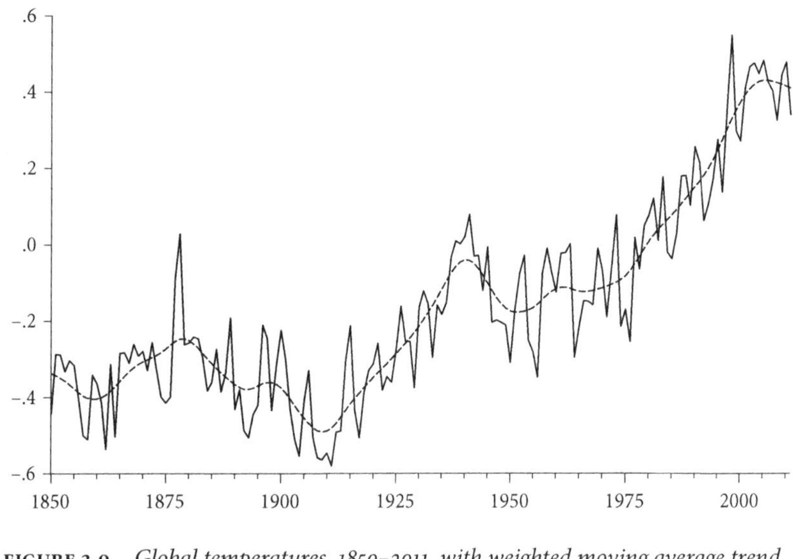

FIGURE 3.9 *Global temperatures, 1850–2011, with weighted moving average trend superimposed*

Of course, an instrumental temperature record like this is only available for a limited time in the past, but several temperature reconstructions, using proxies such as tree ring and borehole data, have taken the record back some 2000 years. This has provoked great interest and debate, and it is probably fair to say that the 'statistical jury' is currently still out on whether there is incontrovertible evidence that the recent warming trend is 'uniquely anthropogenically forced' or whether it may be part of temperature's 'natural variability'.[5]

3.5 Decomposing a time series

The moving averages fitted in the above examples have been interpreted as *trends*, the long-run, smoothly evolving component of a time series. In general, an observed time series may be *decomposed* into several components. We will consider a three component decomposition in which the observed series X_t is decomposed into trend, T_t, seasonal, S_t, and irregular, I_t, components. The decomposition can either be *additive*

$$X_t = T_t + S_t + I_t \tag{3.15}$$

or *multiplicative*

$$X_t = T_t \times S_t \times I_t \tag{3.16}$$

although this distinction is in a sense artificial, as taking logarithms of (3.16) produces an additive decomposition for $\ln X_t$. The *seasonal* component is a regular, short- term, annual cycle, so that it can only appear in series observed at higher than an annual frequency, typically monthly or quarterly. Since it is a regular cycle, it should be relatively easy to isolate. The *irregular* component is what is left over after the trend and seasonal components have been removed. It therefore should be random and hence unpredictable.

The *seasonally adjusted* series is then defined as either

$$X_t^{SA} = X_t - S_t = T_t + I_t$$

or

$$X_t^{SA} = \frac{X_t}{S_t} = T_t \times I_t$$

Other components are sometimes considered. With macroeconomic series such as GDP, a *business cycle* component is often included: we assume here that this cycle is part of the trend component. With sales data, there can also be a *trading day* component, where the irregular needs adjusting for, say, the number of trading days, weekends or bank holidays in a month.

Decomposing and seasonally adjusting UK retail sales

These ideas, and how the components may be computed, are illustrated using quarterly data on the volume of UK retail sales from 1986 to 2011, which is plotted in Figure 3.10.

Retail sales are seen to have a generally upward trend, with pronounced seasonal variations about this trend, which have increased in amplitude (range) over time. In such circumstances, a multiplicative decomposition is appropriate, but, as noted above, if we take logarithms of (3.16) we obtain the additive decomposition (3.15) for $\ln X_t$ rather than X_t.

To obtain the trend component, we use the centred MA(4) (3.14), which is shown superimposed on the logged series in Figure 3.11(a).

This is indeed a smooth series, but does not have a consistent upward trend: retail sales were very 'flat' in the recession of the early 1990s, and again in the last two years of the sample.

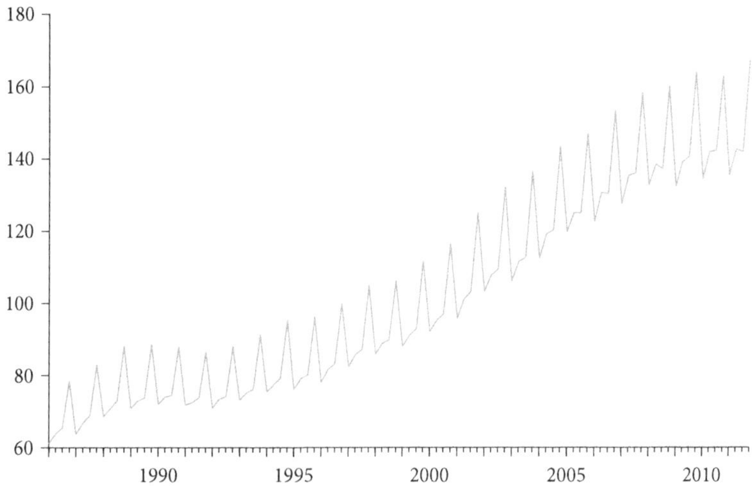

FIGURE 3.10 *Volume of UK retail sales, quarterly 1986–2011*

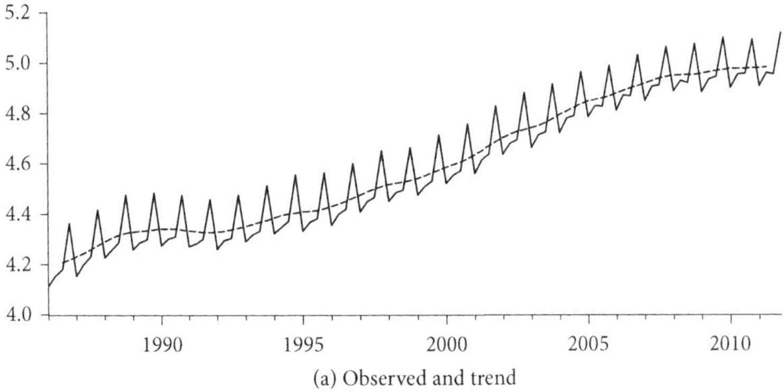

FIGURE 3.11(a) *Decomposition of the logarithm of UK retail sales, 1986 Q1–2011 Q4; Observed and trend*

To isolate the seasonal component, we first subtract the trend from the observed series:

$$\ln X_t - T_t = S_t + I_t$$

This 'trend-free' series is the sum of the seasonal and irregular components, which somehow need to be disentangled. We can do this by making the 'identifying' assumption that I_t should, on average, be equal to zero (if it was not, then a portion of it would be predictable and should be part of either the trend or the seasonal). This allows S_t to be calculated by taking the average of each quarter across years. Thus, for example,

$$S_t(Q1) = \frac{\ln X_{1986Q1} + \ln X_{1987Q1} + \ldots + \ln X_{2011Q1}}{26} = -0.069$$

and the other seasonal *factors* are calculated to be

$$S_t(Q2) = -0.035 \quad S_t(Q3) = -0.031 \quad S_t(Q4) = 0.135$$

These factors are required to sum to zero, and so would need adjusting if the raw calculations lead to a non-zero sum (if this sum is $a \neq 0$, say, then $a/4$ should be subtracted from each factor). We see that the fourth quarter of each year has a large positive seasonal, obviously due to the run-up to Christmas – always a crucial period for retailers – and this is compensated by smaller negative factors for the other three quarters. The seasonal pattern is shown in Figure 3.11(b): this method forces the seasonality to be 'constant' over time, but more sophisticated seasonal adjustment procedures allow seasonal patterns to evolve.[6]

The irregular is now calculated 'by residual' as

$$I_t = \ln X_t - T_t - S_t$$

It is plotted in Figure 3.11(c). and turns out to be small compared to the other components and clearly random.[7]

The seasonally adjusted series

$$\ln X_t^{SA} = \ln X_t - S_t = T_t + I_t$$

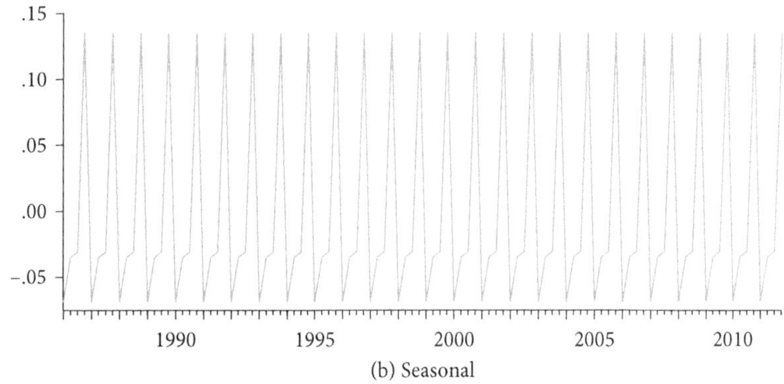

FIGURE 3.11(b) *Decomposition of the logarithm of UK retail sales; 1986 Q1–2011 Q4; Seasonal*

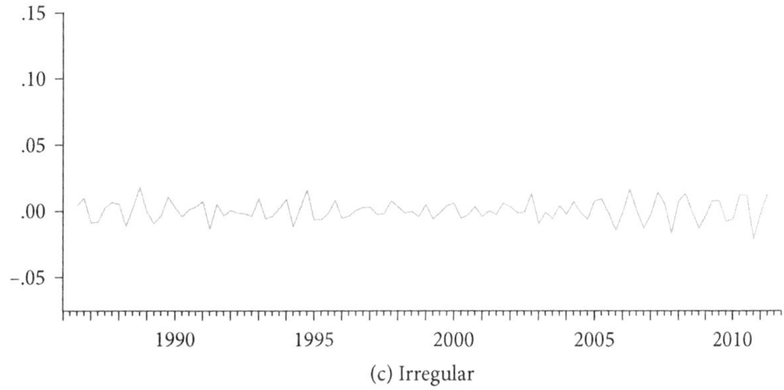

FIGURE 3.11(c) *Decomposition of the logarithm of UK retail sales; 1986 Q1–2011 Q4; Irregular*

is shown with the unadjusted series in Figure 3.12.

From its definition, the seasonally adjusted series is the trend plus the irregular component. Since the irregular is small, it just adds minor random fluctuations to the smooth trend.

The annual growth rate of seasonally adjusted retail sales is shown in Figure 3.13.

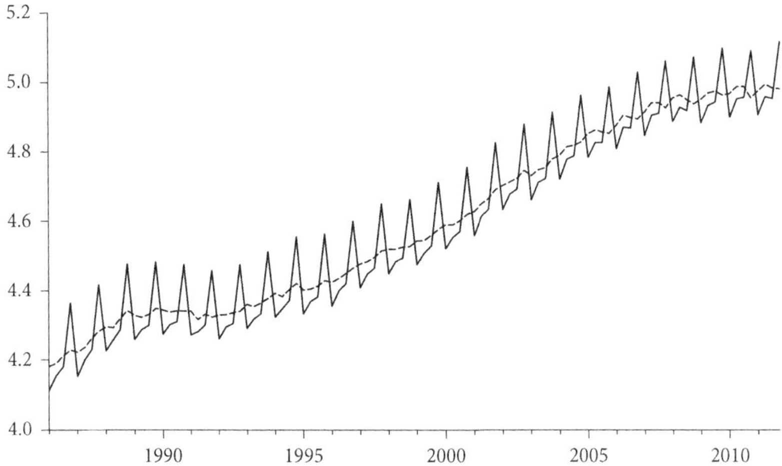

FIGURE 3.12 *Observed and seasonally adjusted UK retail sales*

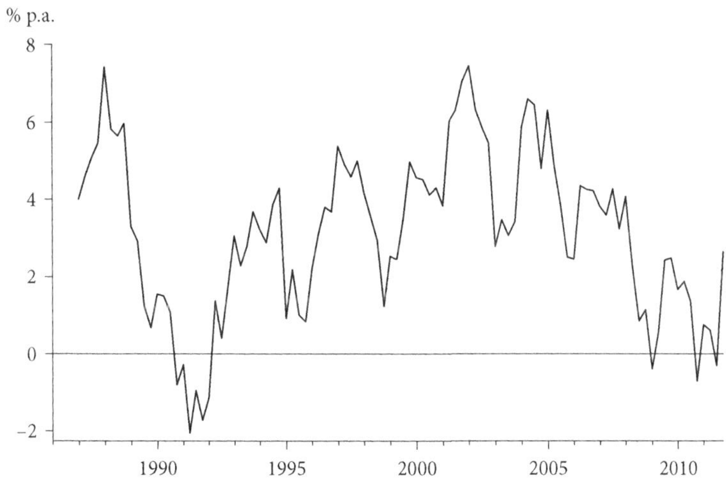

FIGURE 3.13 *Annual growth rate of seasonally adjusted retail sales*

Growth is generally between 2 and 8% per annum, except for the two recessionary periods of the early 1990s and the late 2000s.

Notes

1 While logarithms to any base can be taken, it is conventional to use logs to the base e ($\log_e$, sometimes known as *Napierian logarithms*). Rather than $\log_e$, however, the slightly simpler 'ln' will be used to denote such logarithms. Equation (3.2) uses the standard rules of logarithms: $\ln xy = \ln x + \ln y$ and $\ln x^a = a\ln x$.

2 A.W. Phillips, 'The relation between unemployment and the rate of change of money-wage rates in the UK, 1861–1957', *Economica* 25 (1958), 283–299. The original form of the Phillips curve used here has long been regarded as being far too simplistic and has been extended in various ways: see, for example, the 'New Keynesian Phillips curve' of Olivier Blanchard and Jordi Gali, 'Real wage rigidities and the New Keynesian model', *Journal of Money, Credit and Banking* 39 (2007), 35–65. A retrospective look at the Phillips curve's place in macroeconomics is provided by *Understanding Inflation and the Implications for Monetary Policy: A Phillips Curve Retrospective* (MIT Press, 2009).

3 A useful introductory reference is Terence C. Mills, *Modelling Trends and Cycles in Economic Time Series* (Palgrave Macmillan, 2003).

4 The original reference to the Hodrick–Prescott filter, often referred to as the H–P filter, is Robert J. Hodrick and Edward C. Prescott, 'Postwar US business cycles: an empirical investigation', *Journal of Money, Credit and Banking* 19 (1997), 1–16.

5 For detailed analyses of current temperature trends, see Terence C. Mills, 'Modelling current trends in Northern Hemisphere temperatures', *International Journal of Climatology* 26 (2006), 867–884; 'Modelling current temperature trends', *Journal of Data Science* 7 (2009), 89–97; 'Skinning a cat: stochastic models for assessing temperature trends', *Climatic Change* 10 (2010), 415–426; and 'Is global warming real? Analysis of structural time series models of global and hemispheric temperatures', *Journal of Cosmology* 8 (2010), 1947–1954.

6 A widely used seasonal adjustment method has been developed by the US Bureau of the Census and is known as X-11. Its latest version is available within *EViews*.

7 There is a suggestion that the irregular has begun to show signs of having a 'seasonal' pattern since the mid-2000s. This might be an indication that the seasonal component may not be completely deterministic but may have started to evolve slowly.

4
Index Numbers

Abstract: *Transforming data into index form is proposed, along with chain-linking an index. More sophisticated index numbers, such as the Laspeyres and Paasche, are developed through an example in which an energy price index is constructed. A short discussion is provided on how the RPI and CPI indices for the UK are calculated. The links between price, quantity and expenditure indices are investigated, and this leads to a discussion of how nominal data can be deflated to obtain real data, with a variety of examples being provided, such as how to define real interest and exchange rates and how to compute real petrol prices. The Lorenz curve and the Gini coefficient are introduced as indices for measuring income inequality.*

4.1 Transforming data to index form

Data in the form of *index numbers* are ubiquitous throughout applied economics, appearing in many forms and in many areas, so that two obvious questions to ask are: *what exactly* are index numbers and *why* are they used?

An index number can be thought of as a type of summary statistic, so that it summarises a large amount of information on, say, prices and quantities, in a single value. It is thus used when the underlying raw data is too time-consuming to present, or too complicated to comprehend easily.

A simple example: UK GDP

Table 4.1 shows, in column (1), UK GDP at 2008 prices (that is, a volume measure, typically referred to as *output*) for the first decade of the 21st century.

Because it is measured in £ million, the scale is hard to interpret, particularly when trying to assess how quickly the values are growing. We can simply convert the observations to *index form*, however, by picking a *base year*, arbitrarily converting the base year value to a simple to understand number, known as the *base*, and then adjusting all the other values to the same base. Column (2) does this by defining 2005 to be the base year, with a base of 100. Each year's GDP is then converted by the formula

TABLE 4.1 *UK GDP at 2008 prices, £ million, 2000–2010*

	Income (1)	Index (2)
2000	1185305	86.8
2001	1222650	89.5
2002	1255142	91.9
2003	1299381	95.1
2004	1337782	98.0
2005	1365685	100.0
2006	1401290	102.6
2007	1449861	106.2
2008	1433871	104.6
2009	1371163	100.4
2010	1399850	102.5

$$Index_t = 100 \times \frac{Observed_t}{Observed_{base}}$$

Thus

$$Index_{2000} = 100 \times \frac{1185305}{1365685} = 86.8, \text{ etc.}$$

The output observations are thus measured relative to the base year. Note that any year can be chosen to be the base year, and any value as the base, although it makes sense to pick a simple value. We may then easily calculate that the 'output loss' during the global financial crisis between 2007 and 2009 was 5.5%: $(100 \times (100.4 - 106.2)/106.2 = -5.5)$.

Many economic time series are reported in index form in official publications. One problem that occasionally occurs, particularly when taking data from different 'vintages' of publications, is that the base changes every so often, usually after five or ten years, so that the data 'jumps' at these years. This can easily be solved by *chaining*, to obtain a *chain-linked* index.

Consider Table 4.2, where we now report GDP from 1991 up until 2010.

TABLE 4.2 *Chain-linked GDP indices*

	Base 2005 (1)	Base 1995 (2)	Chained 2005 (3)	Chained 1995 (4)
1991		90.9	66.0	90.9
1992		91.0	66.1	91.0
1993		93.1	67.6	93.1
1994		97.0	70.4	97.0
1995		100.0	72.6	100.0
1996		102.9	74.7	102.9
1997		106.4	77.2	106.4
1998		110.5	80.2	110.5
1999		114.5	83.1	114.5
2000	86.8	119.6	86.8	119.6
2001	89.5		89.5	123.2
2002	91.9		91.9	126.5
2003	95.1		95.1	131.0
2004	98.0		98.0	134.9
2005	100.0		100.0	137.7
2006	102.6		102.6	141.3
2007	106.2		106.2	146.2
2008	104.6		104.6	144.0
2009	100.4		100.4	138.3
2010	102.5		102.5	141.1

Unfortunately, the earlier data has a base year of 1995, so that the index appears to drop from 119.6 in 2000 (see column (2)) to 89.5 in 2001 (see column (1)). This ~25% drop in a single year is clearly the result of trying to match up data with two different bases, and so the artificial decrease must be eradicated. This can always be done so long as we have an 'overlapping' observation, that is, one date at which we have observations from both bases. In our example, 2000 is the overlapping observation, taking the value 86.8 on the 2005 base, and 119.6 on the 1995 base. Clearly, GDP in 2000 must be unaffected by the choice of base year, so that, if we wish to stick to the 2005 base, the observations in column (2) must be multiplied by the *conversion factor* of $(86.8/119.6) = 0.726$, giving column (3).

Alternatively, if we want to use the 1995 base, the observations in column (1) need to be *divided* by the conversion factor (or multiplied by its inverse, $1/0.726 = 1.377$), thus giving column (4). In both cases, the jump in the data at 2000 has been smoothed out. If more than one base change occurs, then chaining can be used repeatedly.

While index numbers of this type are clearly simple, they have to be interpreted carefully. For an individual series, an index number in isolation is meaningless; what does a GDP index of 102.5 in 2010 actually tell you? It only becomes meaningful if it is compared to another value: in 2005 the index was 100, so that we can say that GDP has grown by 2.5% over the period 2005 to 2010. Comparing GDP indices across countries, for example, will be misleading unless the base year, base *and* underlying GDP values in the base year are identical – which is unlikely, to say the least!

4.2 More sophisticated indices

This GDP example illustrates the transformation of an observed data set into index form: essentially, all we have done is to scale the original data so that it should be easier to interpret. More sophisticated index numbers attempt to combine information on a set of prices, say, by weighting the individual prices by some measure of their relative importance (perhaps given by the quantities sold of each of the products). This would define a *price index*, of which the Consumer Price and Retail Price Indices (CPI and RPI respectively) are the most familiar in the UK, although the FTSE stock market indices are another popular example.

TABLE 4.3 *UK energy prices, 2000–2010*

	Coal	Gas	Electricity	Petroleum
2000	76.7	73.0	84.4	91.5
2001	80.4	75.0	84.0	86.8
2002	84.5	79.7	84.4	84.0
2003	86.3	81.2	85.3	87.1
2004	90.8	87.1	90.4	91.8
2005	100.0	100.0	100.0	100.0
2006	107.5	131.9	121.7	105.5
2007	115.2	142.1	131.4	108.4
2008	137.2	170.1	151.9	124.7
2009	161.3	193.5	158.8	114.7
2010	161.3	182.0	154.9	134.1

Constructing an energy price index

Consider the data, shown in Table 4.3, on UK energy prices for the recent past.[1] Here we have four categories of fuel, coal (and coke), gas, electricity and petroleum, and we wish to construct an overall energy price index for each year.

Taking an unweighted average (e.g., calculating an index for 2000 as the price average $(76.7 + 73.0 + 84.4 + 91.5/4 = 81.4)$ would be unwarranted unless each fuel category was equally important. If we measure 'importance' by the quantities used of each fuel, then we can construct a weighted average using these quantities as weights.

But what quantities do we choose? A traditional approach, similar to the previous example, is to choose a base year and to use the quantities prevailing in that year as a set of *base-year weights*. Suppose we choose 2000 as our base year, for which the quantities were

Coal	3.581
Gas	57.077
Electricity	28.325
Petroleum	66.293

The cost of the 'basket' of energy in 2000 is then calculated by summing the costs of 'purchasing' each of the year 2000 amounts of energy consumed, as follows:

	2000 Price	2000 Quantity	Price × Quantity
Coal	76.7	3.581	274.66
Gas	73.0	57.077	4166.62
Electricity	84.4	28.325	2390.63
Petroleum	91.5	66.293	6065.81
Total cost			12897.72

We now need to find what the 2000 basket of energy would cost in 2001. This calculation is

	2001 Price	2000 Quantity	Price × Quantity
Coal	80.4	3.581	287.92
Gas	75.0	57.077	4280.77
Electricity	84.0	28.325	2379.30
Petroleum	86.8	66.293	5754.23
Total			12702.22

Similar calculations for the years 2002 to 2010 lead to the set of costs

2000	12897.72
2001	12702.22
2002	12810.87
2003	13133.94
2004	13942.84
2005	15527.60
2006	18354.48
2007	19431.24
2008	22769.42
2009	23723.83
2010	27243.06

The index of energy prices can then be calculated by taking the ratio of the costs for each year relative to 2000 (and conventionally multiplying by 100). This leads to what is known as a *Laspeyres price index*:

	Index	% change
2000	100.00	–
2001	98.48	−1.52
2002	99.33	0.86
2003	101.83	2.52
2004	108.10	6.16
2005	120.39	11.37
2006	142.31	18.21
2007	150.66	5.87
2008	176.54	17.18
2009	183.94	4.19
2010	187.96	2.19

The Laspeyres index shows that energy prices dropped in 2001 compared to 2000 before increasing slowly in 2002 and 2003 and then much faster until 2008, since when the rate of increase has declined.

Mathematically, the Laspeyres price index is defined as

$$P_t^L = 100 \times \frac{\sum_i p_{i,t} q_{i,0}}{\sum_i p_{i,0} q_{i,0}} \tag{4.1}$$

In this formula, $p_{i,0}$ and $q_{i,0}$ are the base year prices and quantities, and the $p_{i,t}$ are the 'current' year prices.

Rewriting (4.1) as

$$P_t^L = 100 \times \frac{\sum_i \left(\frac{p_{i,t}}{p_{i,0}} \right) p_{i,0} q_{i,0}}{\sum_i p_{i,0} q_{i,0}} = 100 \times \frac{\sum_i \left(\frac{p_{i,t}}{p_{i,0}} \right) w_{i,0}^L}{\sum_i w_{i,0}^L} \tag{4.2}$$

where $w_{i,0}^L = p_{i,0} q_{i,0}$, shows that P_t^L is a weighted average of the *price relatives* $p_{i,t}/p_{i,0}$, with the weights given by the base year *expenditures* $p_{i,0}\, q_{i,0}$.

Going a step further and defining the *expenditure share* as

$$s_{i,0} = \frac{p_{i,0} q_{i,0}}{\sum_i p_{i,0} q_{i,0}} = \frac{w_{i,0}^L}{\sum_i w_{i,0}^L}$$

allows the index to be written as

$$P_t^L = 100 \times \sum_i \left(\frac{p_{i,t}}{p_{i,0}} \right) s_{i,0}$$

The choice of 2000 as the base year was arbitrary. If another year was chosen, then the base year expenditures and the price relatives in (4.2) would alter and so would the value of the index.

There is a related, probably more significant, defect with the Laspeyres index. As relative prices alter over time, one would expect quantities consumed to change. This is not allowed for in the index, which is calculated using unchanging base year quantities. These may therefore become unrepresentative after a while, and the index then has to be *rebased* using quantities of a more recent vintage. We can see these changes occurring in the quantities of energy consumed over successive years in Table 4.4 (particularly coal, whose use has declined by more than 40% during the decade).

TABLE 4.4 *Quantities of energy consumed in the UK, 2000–2010*

	Coal	Gas	Electricity	Petroleum
2000	3.581	57.077	28.325	66.293
2001	3.470	57.814	28.609	67.084
2002	2.946	55.234	28.667	66.099
2003	2.758	56.701	28.910	66.772
2004	2.583	57.080	29.144	68.647
2005	2.254	55.384	29.981	69.449
2006	2.131	52.633	29.684	69.837
2007	2.312	49.961	29.463	69.488
2008	2.297	51.064	29.421	67.290
2009	2.072	45.799	27.749	63.962
2010	2.008	51.607	28.230	63.794

An alternative to using base year weights is to use *current year* weights, thus defining the *Paasche price index* as

$$P_t^P = 100 \times \frac{\sum_i p_{i,t} q_{i,t}}{\sum_i p_{i,0} q_{i,t}}$$

$$= 100 \times \frac{\sum_i \left(\frac{p_{i,t}}{p_{i,0}}\right) p_{i,0} q_{i,t}}{\sum_i p_{i,0} q_{i,t}} = 100 \times \frac{\sum_i \left(\frac{p_{i,t}}{p_{i,0}}\right) w_{i,t}^P}{\sum_i w_{i,t}^P} \tag{4.3}$$

Here the weights are $w_{i,t}^P = p_{i,0} q_{i,t}$ and the expenditure share form of the index is

$$P_t^P = 100 \times \frac{1}{\sum_i \left(\frac{p_{i,0}}{p_{i,t}}\right) s_{i,t}}$$

The calculation of the Paasche energy price index in 2001 proceeds as

$$P_{2001}^P = 100 \times \frac{(80.4 \times 3.470) + (75.0 \times 57.814) + (84.0 \times 28.609) + (86.8 \times 67.084)}{(76.7 \times 3.470) + (73.0 \times 57.814) + (84.4 \times 28.609) + (91.5 \times 67.084)}$$

$$= 98.48$$

Similar calculations for the later years obtain

	Index	% change
2000	100.00	–
2001	98.48	−1.52
2002	99.19	0.72
2003	101.73	2.56
2004	107.91	6.07
2005	119.89	11.10
2006	140.70	17.36
2007	148.15	5.30
2008	174.28	17.63
2009	179.02	2.72
2010	186.51	4.18

These tend to be slightly lower than their Laspeyres counterparts. This is to be expected if consumption of energy is switched to those fuels that are becoming relatively cheaper, since the Paasche index, by using current weights, can capture this switch.

Both indices have advantages and drawbacks. The Laspeyres is simpler to calculate and to understand, but loses legitimacy over time as its weights become unrepresentative. The Paasche, on the other hand, always has current weights, but is more difficult to calculate (although this is hardly a problem when computers do most of the calculations) and is a little harder to interpret. They can be combined into (*Fisher's*) *ideal index*, which is the *geometric mean* of the two:

$$P_t^I = \sqrt{P_t^L P_t^P} = 100\sqrt{\frac{\sum p_t q_t}{\sum p_0 q_0} \frac{\sum p_t q_0}{\sum p_0 q_t}}$$

Thus, for 2010, the ideal index is

$$P_t^I = \sqrt{187.96 \times 186.51} = 187.23$$

4.3 Construction of the RPI and CPI

Both the RPI and CPI are annually chain-linked indices. Each year a separate index (based at 100 in January) is calculated, and each year's indices are then chained together to produce an index covering several years. Within each year the RPI, say, is a fixed-quantity price index, so that it measures the change in a basket of goods of fixed composition, quantity and, as far as possible, quality. It is thus a 'Laspeyres- type' index of the form

$$RPI_t = 100 \times \frac{\sum_i p_{i,t} q_{i,b}}{\sum_i p_{i,0} q_{i,b}} = 100 \times \frac{\sum_i \left(\frac{p_{i,t}}{p_{i,0}}\right) w_{i,b}}{\sum_i w_{ib}}$$

where $w_{i,b} = p_{i,0} q_{i,b}$. This index differs from (4.1) in that $q_{i,b}$ is used rather than $q_{i,0}$, so that quantities from a base period b are used rather than those from base period 0. This is because it is, in practice, impossible to get period 0 quantities accurately, and these are therefore calculated from data available from the most recent 12 months. Of course, the exact construction of the RPI and CPI is a good deal more complicated, but this is the general principle underlying their construction.[2]

4.4 Price, quantity and expenditure indices

Just as we can calculate price indices, it is also possible to calculate *quantity* and *expenditure* (or *value*) indices. The Laspeyres and Paasche quantity indices are defined as

$$Q_t^L = 100 \times \frac{\sum_i p_{i,0} q_{i,t}}{\sum_i p_{i,0} q_{i,0}} \tag{4.4}$$

and

$$Q_t^P = 100 \times \frac{\sum_i p_{i,t} q_{i,t}}{\sum_i p_{i,t} q_{i,0}} \tag{4.5}$$

while the expenditure index is

$$E_t = \frac{\sum_i p_{i,t} q_{i,t}}{\sum_i p_{i,0} q_{i,0}} \tag{4.6}$$

There is an important link between price, quantity and expenditure indices. Just as multiplying the price of a single good by the quantity purchased gives the total expenditure on the good, so the same is true of index numbers. Or, to put it another way, an expenditure index can be decomposed as the product of a price index and a quantity index. However, the decomposition is both subtle and non-unique, as the following pair of equations show:

$$E_t = \frac{\sum_i p_{i,t}q_{i,t}}{\sum_i p_{i,0}q_{i,0}} = \frac{\sum_i p_{i,t}q_{i,t}}{\sum_i p_{i,t}q_{i,0}} \times \frac{\sum_i p_{i,t}q_{i,0}}{\sum_i p_{i,0}q_{i,0}} = Q_t^P \times P_t^L$$

$$E_t = \frac{\sum_i p_{i,t}q_{i,t}}{\sum_i p_{i,0}q_{i,0}} = \frac{\sum_i p_{i,t}q_{i,t}}{\sum_i p_{i,0}q_{i,t}} \times \frac{\sum_i p_{i,0}q_{i,t}}{\sum_i p_{i,0}q_{i,0}} = P_t^P \times Q_t^L$$

Thus the expenditure index is *either* the product of a Laspeyres price index and a Paasche quantity index *or* the product of a Paasche price index and a Laspeyres quantity index. Thus two decompositions are possible and will give slightly different results.

It is also evident that a quantity index can be constructed by dividing the expenditure index by a price index, since

$$Q_t^P = \frac{E_t}{P_t^L} \quad \text{and} \quad Q_t^L = \frac{E_t}{P_t^P}$$

This is often the easiest way of computing quantity indices. With the available energy data, in particular the price indices, we can make the following calculations

	Σpq	E	Q^L	Q^P
2000	12897.72	100.00	100.00	100.00
2001	12841.09	99.56	101.10	101.09
2002	12622.90	97.87	98.67	98.53
2003	13123.83	101.75	100.02	99.92
2004	14142.62	109.65	101.61	101.43
2005	15706.80	121.78	101.58	101.15
2006	18151.72	140.74	100.03	98.90
2007	18769.74	145.53	98.23	96.60
2008	21861.25	169.50	97.26	96.01
2009	20939.30	162.35	90.69	88.26
2010	22643.97	175.57	94.13	93.40

Thus expenditure on energy fell in 2001 and 2002 from its 2000 value before rebounding in 2003 and then increasing quickly until 2008, after which it declined in 2009 and then increased again in 2010. The quantity indices, however, show increases in 2001 which almost offset the decline in the price indices. The quantity indices then remained relatively flat before declining throughout 2006–2009, although there was again a rebound in 2010. The quantity indices were much more stable than the price indices, which increased rather rapidly: increases in energy expenditure were thus entirely a consequence of increases in energy prices.

4.5 Deflating nominal data to obtain real data

Constructing a quantity index by dividing an expenditure index by a price index is known as *deflating*. It can be used in a more general context. Expenditure indices are an example of a *nominal series* – macroeconomic examples are nominal GDP, nominal consumption and nominal money supply. By dividing a nominal series by a price index, such as the RPI or the CPI, we obtain a *real series* (for example, real GDP – generally known as output – real consumption or real money). When inflation is generally positive, 'stripping out' the price component of a nominal series will lead to a real series having a lower growth rate than its nominal counterpart. When plotted together, the nominal version of a variable will then have a steepening slope when compared to its real counterpart.

Nominal and real GDP

This feature is illustrated in Figure 4.1, which plots nominal (i.e., at market prices) and real (that is, at constant, in this case 2008, prices) GDP for the UK annually from 1948 to 2010.

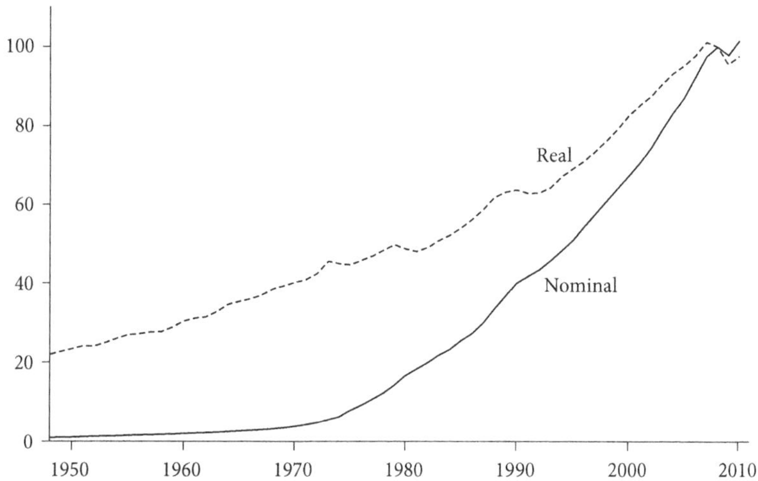

FIGURE 4.1 *UK GDP, annual 1948–2010*

Real GDP is obtained by deflating the nominal series by a price index, the GDP deflator. The series are in index form with 2008 as the base year: both thus equal 100 in that year, which is the 'cross-over' point. Nominal GDP starts at a much lower level than its real counterpart and finishes higher, so that it must have grown faster. Average growth rates for the two series are 7.7% for nominal GDP and 2.4% for real GDP, which implies that the GDP deflator must have grown on average by 5.3%, which is thus average GDP inflation. Interestingly, the average growth of real GDP is close to what the Treasury has long regarded as trend growth for output, 2.5%.

Nominal and real interest rates

A different application of deflation is to interest rates. The *nominal* interest rate earned on an asset makes no allowance for *expected inflation* over the period for which the asset is held, which will decrease the purchasing power of the interest earned on the asset. In general, the *real interest rate*, r_t, can be defined as

$$r_t = i_t - \pi_t^e$$

where i_t is the nominal interest rate and π_t^e is expected inflation. Expected inflation over the future holding period is, of course, unobserved, and there are many schemes for estimating it. The simplest is to assume that expected inflation equals current inflation, π_t, and thus calculate r_t as the difference between i_t and π_t.

Figure 4.2 shows nominal and real interest rates for the UK for the data originally used in Figure 2.6, although only observations from 1925 are plotted for clarity and topicality.

Except for the decade up to the mid-1930s, real rates have always been lower than nominal rates (apart from 2009, when inflation was negative), and for short periods have even been negative, when inflation was greater than nominal interest rates. The period up to 1935 corresponds to the inter-war years of depression, when prices were falling, so that inflation was negative and real rates were higher than nominal rates.

Nominal and real exchange rates

In Figure 3.8 we plotted the \$–£ exchange rate. This is a nominal exchange rate, measuring the *foreign price of domestic currency*, that is, how many dollars are required to purchase £1. Similarly, the €–£ exchange rate

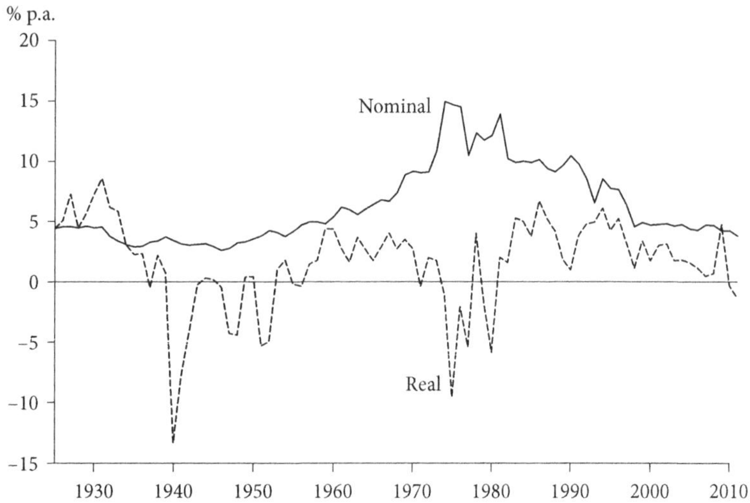

FIGURE 4.2 *Nominal and real UK interest rates, 1925–2011*

shows how many euros are required to purchase £1. Exchange rates in this form, which we shall denote as *e*, are popular in the UK, but they could be (and often are) defined as the *domestic price of foreign currency*, which would give us £–$ and £–€ exchange rates denoted as the *reciprocal* of *e*, written as e^*, which equals 1/*e*. An increase in *e* constitutes an *appreciation* of the domestic currency (so if the $–£ rate goes up from 1.8 to 2, then the £ sterling has appreciated), while an increase in e^* constitutes a *depreciation* (so if the £–$ rate goes up from 0.5 to 0.6 then the £ sterling has depreciated, since the $–£ rate has gone down, from 2 to 1.67). It is thus vital that users of exchange rate data are aware of which definition is being used in a particular context.

The *real exchange rate* measures a country's competitiveness in international trade, and is given by the ratio of goods prices abroad, measured in domestic currency, relative to the price of the same goods at home. Thus, using the superscript *R* to denote a real rate, we have either

$$e^R = e\frac{P^f}{P^d}$$

or

$$e^{*R} = e^*\frac{P^d}{P^f}$$

where P^d and P^f are the domestic and foreign price levels.

As an example of real exchange rate calculations, consider the following (annual average) data for the UK and the US.

	e: \$–£	e^*: £–\$	P^{UK}	P^{US}	e^R	e^{*R}
1998	1.657	0.604	100.0	100.0	1.657	0.604
2011	1.603	0.624	142.1	138.0	1.557	0.642

Between 1998 and 2011, sterling depreciated nominally by 3.3%, but because prices rose more in the UK than in the US over this period the real rate depreciated by 6.0%, thus making UK goods more competitive than might be thought from just the movement in the nominal exchange rate.

Real petrol prices

The price of petrol has consistently vexed motorists over the years, particularly those in the UK, where approximately 60% of the price (132.9 pence per litre in 2011, a record high) is now taken as government taxation.

How has the real price of petrol moved over the years? Figure 4.3 shows the real price of a litre of petrol in the UK compared to the 2011 price (the base year is 2011 and the base is the actual price for that year) since the beginning of the 'motoring age' in the early part of the 20th century.

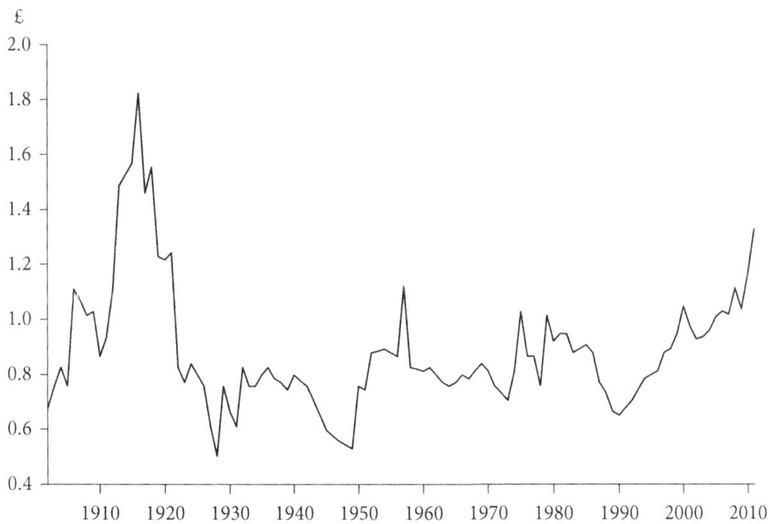

FIGURE 4.3 *Real price of a litre of petrol compared to 2011*

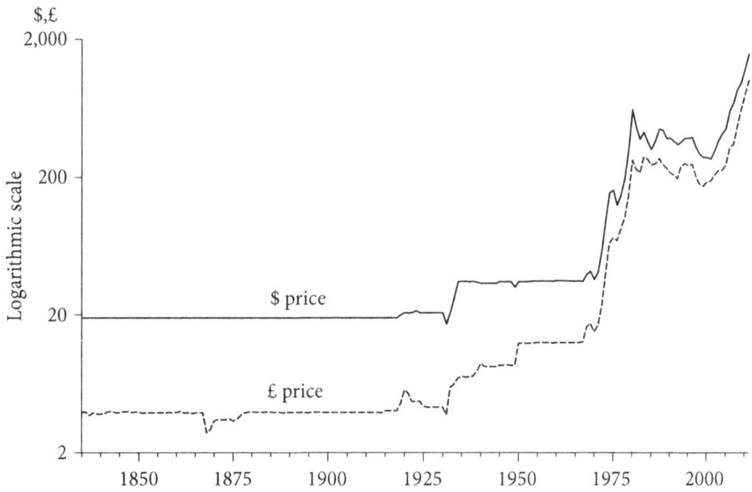

FIGURE 4.4 *Dollar and sterling price of an ounce of gold, 1835–2011*

The 2011 price was exceeded by the real price during World War I (actually from 1913 to 1918), reaching a maximum of 182.5 pence in 1916, but since then the price has always been lower, sometimes considerably so. The minimum price of 50 pence was reached in 1928, during the depths of the inter-war recession, but even during World War II and its aftermath of petrol rationing, prices remained relatively low.

Real sterling gold prices

The price of gold is conventionally given in US dollars per ounce, with gold traditionally being seen as a 'safe haven' in times of political and economic uncertainty and generally as a hedge against inflation and, possibly, exchange rate movements.[3] The price of gold in sterling can easily be calculated by dividing by *e*, the $–£ exchange rate. Figure 4.4 shows the two prices from 1835 until 2011.

The price was effectively fixed until the early 1970s, at which point an unregulated market was allowed to develop and the price adjusted rapidly upwards before stabilising in the early 1980s. Since 2006 the price has again increased substantially, standing at $1571 (£1009) in 2011.

The narrowing differential between the dollar and sterling prices over the period reflects the long-run depreciation of sterling against the dollar. Until 1938, the exchange rate fluctuated in a quite narrow band about $5,

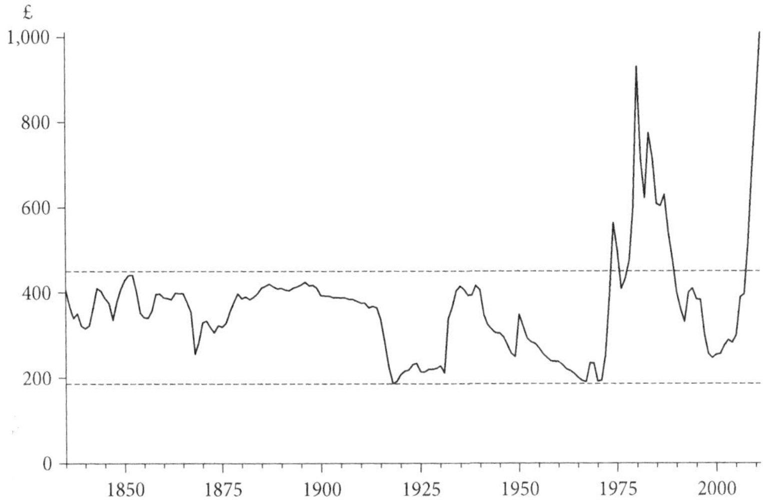

FIGURE 4.5 *Real sterling price of an ounce of gold compared to 2011*

after which successive devaluations and the move to floating exchange rates in the early 1970s has narrowed the difference in the two prices.

What has been the real sterling price of gold? Figure 4.5 shows the sterling price of gold deflated by the RPI, so that prices are measured relative to the 2011 value of £1009.

The 2011 price is seen to be a real as well as a nominal high, eclipsing the previous high of £930 achieved in 1980. However, apart from most of 1974 to 1980 and the years from 2008, which were all periods of high prices, the real sterling price of gold has fluctuated in the range £185 to £450, unlike the nominal price, so that the inflation-hedging properties of gold are clearly identified.

4.6 Inequality indices: the Lorenz curve and the Gini coefficient

The index numbers considered so far are typically used to compare values across time and so become treated as time series data. Another type of index number is used specifically in the measurement of *inequality*, such as inequality in the distribution of income. We have already measured the dispersion of such a distribution using the sample standard deviation, based on the deviation of each observation from the sample mean (recall §2.3). An alternative idea is to measure the difference between *every pair* of observations, and this forms the basis of a statistic known as the *Gini coefficient*. An attractive visual interpretation of this statistic is the *Lorenz curve*, from which the Gini coefficient can easily be calculated. We develop these ideas using the income inequality data set introduced in Table 2.1 and Figure 2.1.

The Lorenz curve

The Lorenz curve plots the cumulative percentage income (on the vertical axis) against the cumulative percentage of countries (on the horizontal axis). The curve takes on the following form:

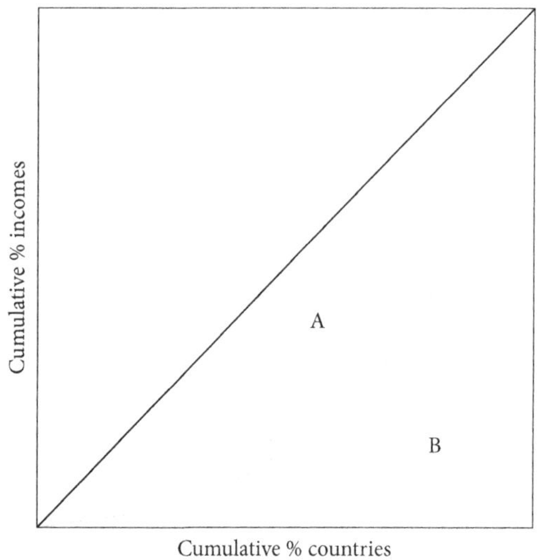

It will be of this shape for the following reasons:

- Since 0% of countries have 0% of income, and 100% of countries have 100% of income, the curve must run from the origin up to the opposite corner.
- Since countries are ranked from the poorest to the richest, the Lorenz curve must lie below the 45° line, which is the line representing complete equality. The further away from the 45° line is the Lorenz curve, the greater is the degree of inequality.
- The Lorenz curve must be concave from above: as we move to the right, we encounter successively richer countries, so that cumulative income grows faster.

The Gini coefficient

The Gini coefficient is a numerical measure of the degree of inequality in a distribution, and can be derived directly from the Lorenz curve. Looking at the schematic form of the curve above, it is defined as the ratio of area A to the sum of areas A and B; that is, if the Gini coefficient is denoted *G* then it is defined as

$$G = \frac{A}{A+B}$$

so that $0 < G < 1$. When there is total equality, the Lorenz curve coincides with the 45° line, area A disappears and $G = 0$. With total inequality (one country having all the income), area B disappears and $G = 1$ (in fact this is only true for an infinite number of countries: with *N* countries the maximum value of *G* is $1-2/(N+1)$). Neither of these two extremes is likely to occur, but in general, the higher is *G*, the greater the degree of inequality.

The Gini coefficient can be calculated from the following formula for area B:

$$B = \frac{1}{2}\left[(x_1 - x_0)\times(y_1 + y_0) + (x_2 - x_1)\times(y_2 + y_1) + \ldots + (x_k - x_{k-1})\times(y_k + y_{k-1})\right]$$

Here the *x* and *y* values are the horizontal and vertical coordinates of the points on the Lorenz curve, with $x_0 = y_0 = 0$ and $x_k = y_k = 100$ being the coordinates of the two end-points; *k* being the number of classes. Area A is then given by

$$A = 5000 - B$$

(This uses the result of the area of a triangle being given by $\frac{1}{2} \times base \times height$. Here, the base and height are both 100, so that the area of the triangle defined by A + B is 5000). Thus

$$G = \frac{5000 - B}{5000}$$

Calculating the Lorenz curve and Gini coefficient for the income inequality data

The calculations required to obtain the two sets of values needed to compute the Lorenz curve for the income inequality data are detailed in Table 4.5, while Figure 4.6 shows the Lorenz curve itself, which plots column (8) of the table against column (6).

TABLE 4.5 *Lorenz curve calculations*

Income class (1)	Class mid-point x (2)	No. of countries f (3)	$f \times x$ (4)	$\%f$ (5)	% Cumulative f (6)	% Income (7)	% Cumulative Income (8)
$0-	1250	52	65000	27.51	27.51	2.58	2.58
$2500-	3750	26	97500	13.76	41.27	3.87	6.46
$5000-	6250	19	118750	10.05	51.32	4.72	11.18
$7500-	8750	16	140000	8.47	59.79	5.56	16.74
$10000-	11250	14	157500	7.41	67.20	6.26	23.00
$12500-	13750	8	110000	4.23	71.43	4.37	27.37
$15000-	16250	5	81250	2.65	74.07	3.23	30.60
$17500-	18750	4	75000	2.12	76.19	2.98	33.58
$20000-	21250	4	85000	2.12	78.31	3.38	36.96
$22500-	23750	6	142500	3.17	81.48	5.66	42.62
$25000-	26250	3	78750	1.59	83.07	3.13	45.75
$27500-	28750	5	143750	2.65	85.72	5.71	51.43
$30000-	31250	5	156250	2.65	88.36	6.21	57.68
$32500-	33750	4	135000	2.12	90.48	5.37	63.04
$35000-	36250	5	181250	2.65	93.13	7.20	70.24
$37500-	38750	1	38750	0.53	93.66	1.54	71.78
$40000-	41250	3	123750	1.59	95.25	4.92	76.70
$42500-	43750	0	0	0	95.25	0	76.70
$45000-	46250	3	138750	1.59	96.84	5.51	82.22
$47500-	48750	1	48750	0.53	97.37	1.94	84.15
$50000-	51250	2	102500	1.06	98.43	4.07	88.23
$52500-	53750	1	53750	0.53	98.96	2.14	90.36
			⋮	⋮	⋮	⋮	⋮
$82500-	83750	1	83750	0.53	99.48	3.33	93.69
⋮	⋮	⋮	⋮	⋮	⋮	⋮	⋮
$157500-	158750	1	158750	0.53	100.00	6.31	100.00
Totals		189	2516250	100.00		100.00	

Notes:
Col (4) = col (2) × col (3)
Col (5) = col (3) ÷ 189
Col (6) = col (5) cumulated
Col (7) = col (4) ÷ 2516250
Col (8) = col (7) cumulated

From the curve it can be seen that the poorest 25% of countries have about 2½% of income, while the richest 10% have 30%. The curve is fairly smooth, and suggests that there is a much greater degree of inequality at the top of the distribution than at the bottom.

Using the data in Table 4.5, the Gini coefficient is obtained by first calculating B as

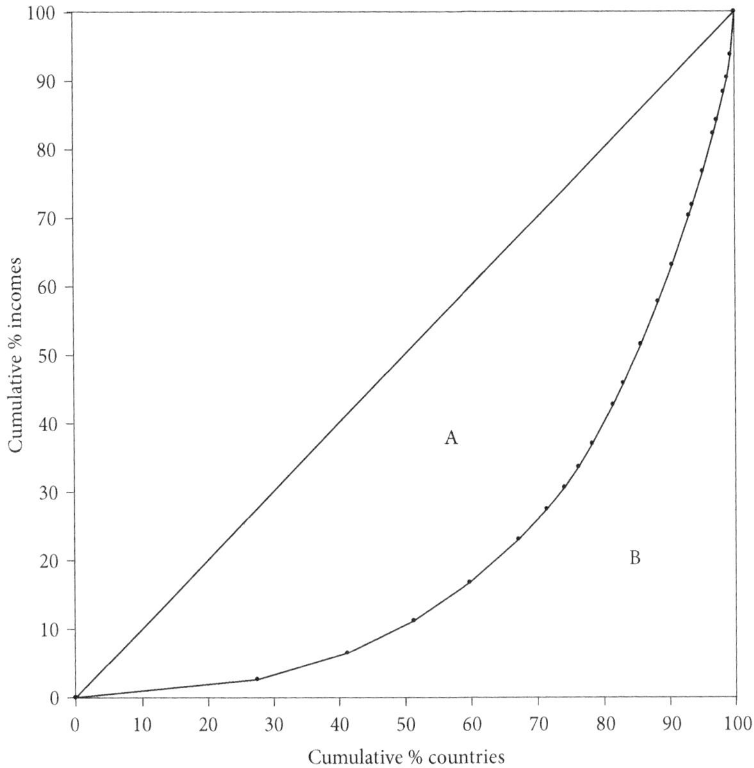

FIGURE 4.6 *Lorenz curve for income equality data*

$$B = \frac{1}{2}\left[\begin{array}{l}(27.51-0)\times(2.58+0)+(41.27-27.51)\times(6.46+2.58)+\ldots\\+(100-99.48)\times(100+93.69)\end{array}\right] = 2079.46$$

Thus

$$G = \frac{5000-2079.46}{5000} = 0.584$$

or approximately 58%. This is consistent with other Gini coefficient estimates of world income, which range between 0.56 and 0.66.

On its own, the Gini coefficient does not tell us very much, but it is useful for looking at inequality movements across countries or over time. The developed European countries and Canada have Gini coefficients in the range of 0.23 (Sweden) to 0.36, while the US Gini index has consistently been over 0.4 since 1980.[4]

Notes

1 The source for this price data and the subsequently introduced energy consumption data is the Office for National Statistics and Department of Energy and Climate Change (DECC): *Quarterly Energy Prices*, table 2.1.1, and *Digest of UK Energy Statistics Annex*, table 1.1.5.

2 Details of the actual construction of the RPI and CPI are given in the *Consumer Price Indices Technical Manual, 2010 edition*, Newport: Office for National Statistics, which is downloadable from www.ons.gov.uk .There has been some debate concerning the calculation of the *elementary aggregates* in the construction of the RPI and CPI: see Duncan Elliott, Robert O'Neill, Jeff Ralph and Ria Sanderson, 'Stochastic and sampling approaches to the choice of elementary aggregate formula', *Office for National Statistics Discussion Paper* (5 October 2012) and *History of and Differences between the Consumer Prices Index and Retail Prices Index* (Office of National Statistics, 2011).

The elementary aggregates are the $p_{i,t}$, which are themselves index numbers of the prices of the 'elementary' goods comprising the *i*-th aggregate: for example, if $p_{i,t}$ is the price of bread, then this is the 'average' price of all bread products; sliced white loaf, wholemeal loaf, bread rolls, ciabatta, pitta ....

3 Discussion and empirical analysis of the role of gold as a hedge, particularly an exchange rate hedge, can be found in Forrest H. Capie, Terence C. Mills and Geoffrey E. Wood, 'Gold as a hedge against the dollar', *Journal of International Financial Markets, Institutions and Money* 15 (2005), 343–352.

Statistical analyses of gold prices are to be found in Terence C. Mills, 'Exploring the relationship between gold and the dollar', *Significance* 1 (2004), 113–115, and 'Statistical analysis of gold price data', *Physica A* 338 (2004), 559–566.

4 See http://en.wikipedia.org/wiki/Gini_coefficient for a useful technical discussion of the Gini coefficient and a listing of the various estimates of the coefficient for a variety of countries.

5
Correlation

Abstract: *Correlation, as a measure of the strength of the relationship between two variables, is introduced by analysing several key economic relationships. After formally defining the correlation coefficient and providing a detailed example of its computation, various pitfalls in using the measure are discussed, as well as the link between correlation and the concept of causality. The possibility that the correlation between two variables could be spurious, it being a consequence of the omission of a third variable related to the other two, is discussed; and the possible presence of spurious correlation is analysed via the calculation of partial correlation coefficients. Some examples examining this possibility are provided.*

5.1 Examining the strength of the relationship between two variables

When discussing the use of transformations in §3.3, we presented an example of the Phillips curve, in which a (non-linear) function was fitted to a scatterplot of inflation and unemployment rates. In discussing the fitted line, it was suggested that the overall fit was 'quite loose, with many observations lying some distance from the line', and that a measure of 'goodness of fit' would be proposed later. We are now going to develop the underlying techniques that will enable such measures to be calculated.

To provide a backdrop to this development, consider the scatterplots shown in Figures 5.1 to 5.3.

Figure 5.1 again shows the scatterplot of inflation and unemployment. Although there is a noticeable 'downward drift' from the top left to the bottom right in the 'cloud' of points, signifying a negative relationship between the two variables (that is, high inflation values are associated with low unemployment and vice versa), the dispersion of the cloud tells us that any such relationship is fairly weak. The superimposed straight

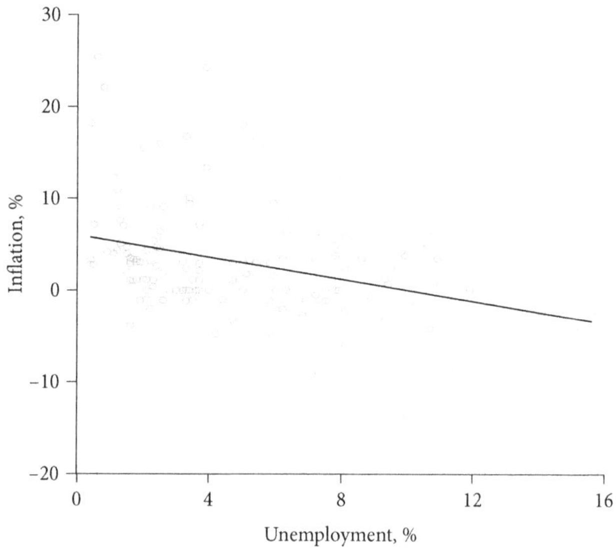

FIGURE 5.1 *Scatterplot of UK inflation and unemployment rates, 1855–2011*

line (the linear line of best fit, whose calculation will be discussed in §6.2) serves to emphasise both the negative relationship and the looseness of the fit.

Figure 5.2, by contrast, shows a very strong positive relationship between real consumption expenditure (*C*) and real income (*Y*) in the UK. Indeed, this strong link between the two variables forms the basis of the fundamental equation in macroeconomic models – the *consumption function*.

Yet another contrast is provided in Figure 5.3, which shows the scatterplot of output (real GDP) growth and unemployment in the UK. The cloud of points is shapeless, suggesting no association between the two variables, and this is confirmed by the fitted line, which is virtually horizontal, signifying that one particular value of output growth (its sample mean, ~2.5%) may be associated with any and every rate of unemployment.

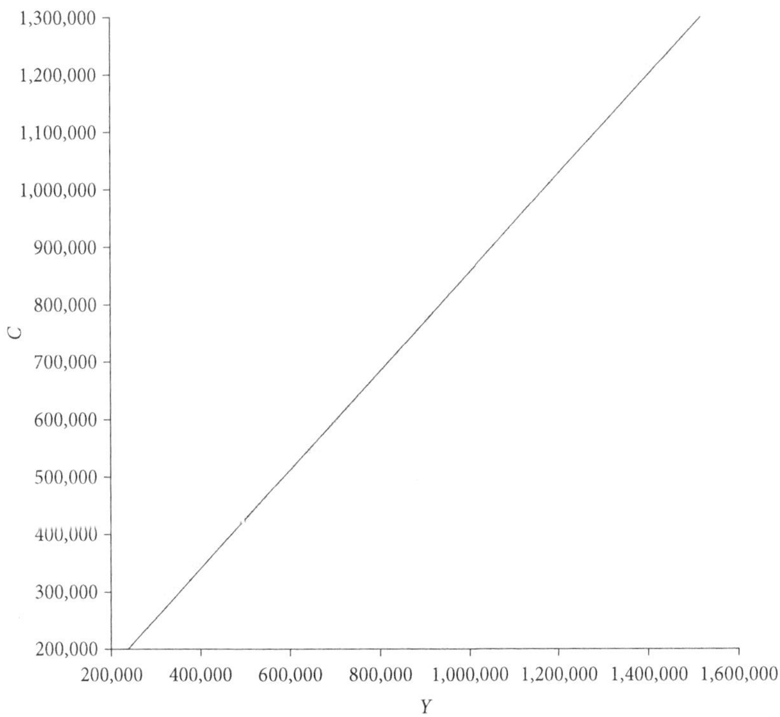

FIGURE 5.2 *Scatterplot of UK real consumption and income, 1948–2010*

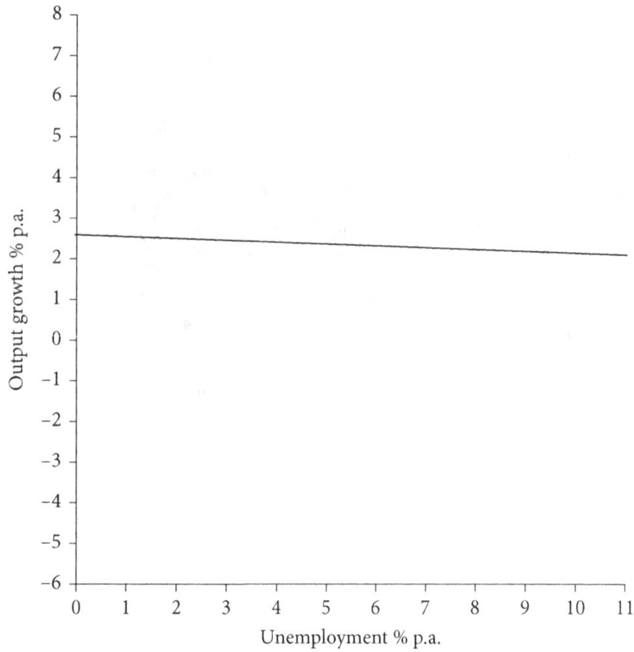

FIGURE 5.3 *Scatterplot of UK output growth and unemployment, 1949–2010*

5.2 The correlation coefficient

While this type of graphical analysis can be quite informative, it is often the case that we would like a single, summary statistic of the strength of the relationship between two variables. This is provided by the (*sample*) *correlation coefficient*, which is defined in the following way. Suppose that we have a sample of N pairs of observations on the variables X and Y

$$(X_1, Y_1), (X_2, Y_2), \ldots, (X_N, Y_N)$$

The correlation coefficient is then given by

$$r_{XY} = \frac{N\sum XY - \sum X \sum Y}{\sqrt{\left(N\sum X^2 - \left(\sum X\right)^2\right)\left(N\sum Y^2 - \left(\sum Y\right)^2\right)}} \qquad (5.1)$$

In (5.1) we have omitted the i subscripts in the summations to avoid cluttering the notation.

How does this formula come about? Recall that the sample variance of X was defined as

$$s_X^2 = \sum\left(X - \bar{X}\right)^2 / (N-1)$$

Similarly, the sample variance of Y is

$$s_Y^2 = \sum\left(Y - \bar{Y}\right)^2 / (N-1)$$

We may also define the *sample covariance* between X and Y to be

$$s_{XY} = \sum\left(X - \bar{X}\right)\left(Y - \bar{Y}\right) / (N-1)$$

This is a measure of how the two variables *covary*, that is, move together. If negative (positive) mean deviations in X are predominantly accompanied by negative (positive) mean deviations in Y, then the product of these mean deviations, and hence s_{XY}, will be positive. If, however, mean deviations of opposite sign predominantly accompany each other, then their product and s_{XY} will be negative. The covariance thus gives a measure of the strength and direction of the covariation existing between X and Y. It has the disadvantage, however, that it will be measured in units that are the product of the units that X and Y are measured in, and hence could take a value of any magnitude which would be almost uninterpretable in most cases. A *scale free* measure can be obtained by dividing the covariance by the square root of the product of the variances, that is, by dividing by the product of the (sample) standard deviations, in which case it can be shown that[1]

$$-1 \leq \frac{s_{XY}}{s_X s_Y} \leq 1$$

In fact, this ratio defines the correlation coefficient, that is:

$$r_{XY} = \frac{s_{XY}}{s_X s_Y} = \frac{\sum\left(X \quad \bar{X}\right)\left(Y - \bar{Y}\right)}{\sqrt{\left(\sum\left(X - \bar{X}\right)^2\right)\left(\sum\left(Y - \bar{Y}\right)^2\right)}} \tag{5.2}$$

Although this looks different to (5.1), the formulae are in fact equivalent, as some tedious algebra would demonstrate![2] Formula (5.1) is useful when the data is available in its 'raw' form, while (5.2) can be used if the data is in mean deviation form. The latter formula will also be easier to use in subsequent algebraic derivations since, on defining the *mean deviations* $x_i = X_i - \bar{X}$ and $y_i = Y_i - \bar{Y}$, it can be written concisely as

$$r_{XY} = \frac{\sum xy}{\sqrt{\sum x^2 \sum y^2}} \tag{5.3}$$

Thus, $r_{XY} > 0$ signifies a positive correlation between X and Y, with $r_{XY} = 1$ signifying a *perfect* positive correlation, where *all* the points in the scatterplot lie *exactly* on an upward sloping (from left to right) straight line. Conversely, $r_{XY} < 0$ signifies a negative correlation, and $r_{XY} = -1$ a perfect negative correlation, where all the points lie on a downward sloping (from left to right) straight line. X and Y are *uncorrelated* if $r_{XY} = 0$, in which case the scatterplot has the appearance of a 'shapeless cloud'.

A simple example: salary, education and experience

Computation of correlations can routinely be done by standard statistical software, but it can be instructive to work through the calculations in a simple example. Table 5.1 shows annual salary Y (in £000) of $N = 12$ employees of a firm, their years of post-school education, X, and years of experience with the firm, Z, along with the data on these variables in mean deviation form. These last columns were calculated using the sample means $\bar{Y} = 30$, $\bar{X} = 5$ and $\bar{Z} = 10$.

Table 5.2 provides the various squares, cross-products and their sums required to compute correlations.

TABLE 5.1 *Salary (Y), education (X) and experience (Z) data for 12 employees*

Employee	Y	X	Z	y	x	z
1	34	5	11	4	0	1
2	24	4	9	−6	−1	−1
3	39	7	12	9	2	2
4	26	5	9	−4	0	−1
5	44	8	13	14	3	3
6	34	4	11	4	−1	1
7	21	1	8	−9	−4	−2
8	24	4	8	−6	−1	−2
9	28	6	10	−2	1	0
10	24	5	8	−6	0	−2
11	34	7	11	4	2	1
12	28	4	10	−2	−1	0
Σ	360	60	120	0	0	0

TABLE 5.2 *Sums of squares and cross-products of the employee data*

Employee	Y^2	X^2	Z^2	XY	ZY	XZ
1	1159	25	121	170	374	55
2	576	16	81	96	216	36
3	1521	49	144	273	468	84
4	676	25	81	130	234	45
5	1936	64	169	352	572	104
6	1156	16	121	136	374	44
7	441	1	64	21	168	8
8	576	16	64	96	192	32
9	784	36	100	168	280	60
10	576	25	64	120	192	40
11	1156	49	121	238	374	77
12	784	16	100	112	280	40
Σ	11338	338	1230	1912	3724	625
Employee	y^2	x^2	z^2	xy	zy	xz
1	16	0	1	0	4	0
2	36	1	1	6	6	1
3	81	4	4	18	18	4
4	16	0	1	0	4	0
5	196	9	9	42	42	9
6	16	1	1	−4	4	−1
7	81	16	4	36	18	8
8	36	1	4	6	12	2
9	4	1	0	−2	0	0
10	36	0	4	0	12	0
11	16	4	1	8	4	2
12	4	1	0	2	0	0
Σ	538	38	30	112	124	25

Using formula (5.1), the three correlations are calculated as

$$r_{XY} = \frac{(12\times 1912)-(60\times 360)}{\sqrt{\left((12\times 338)-60^2\right)\left((12\times 11338)-360^2\right)}} = 0.783$$

$$r_{ZY} = \frac{(12\times 3724)-(120\times 360)}{\sqrt{\left((12\times 1230)-120^2\right)\left((12\times 11338)-360^2\right)}} = 0.976$$

$$r_{XZ} = \frac{(12\times 625)-(60\times 360)}{\sqrt{\left((12\times 338)-60^2\right)\left((12\times 1230)-120^2\right)}} = 0.740$$

while using (5.2), we have

$$r_{XY} = \frac{112}{\sqrt{38\times 538}} = 0.783$$

$$r_{ZY} = \frac{124}{\sqrt{30 \times 538}} = 0.976$$

$$r_{XZ} = \frac{25}{\sqrt{38 \times 30}} = 0.740$$

The three correlations are all positive and high, showing that salary, post-school education and work experience are all strongly correlated, as can also be seen from the scatterplots in Figure 5.4.

These are presented in the form of a 'lower-triangular matrix' showing the scatterplots of the three pairs in an obvious and concise way.

We may now calculate the correlation coefficients for the scatterplots in Figures 5.1–5.3. The inflation–unemployment correlation is –0.34, so that there is a moderately negative correlation between the two variables.

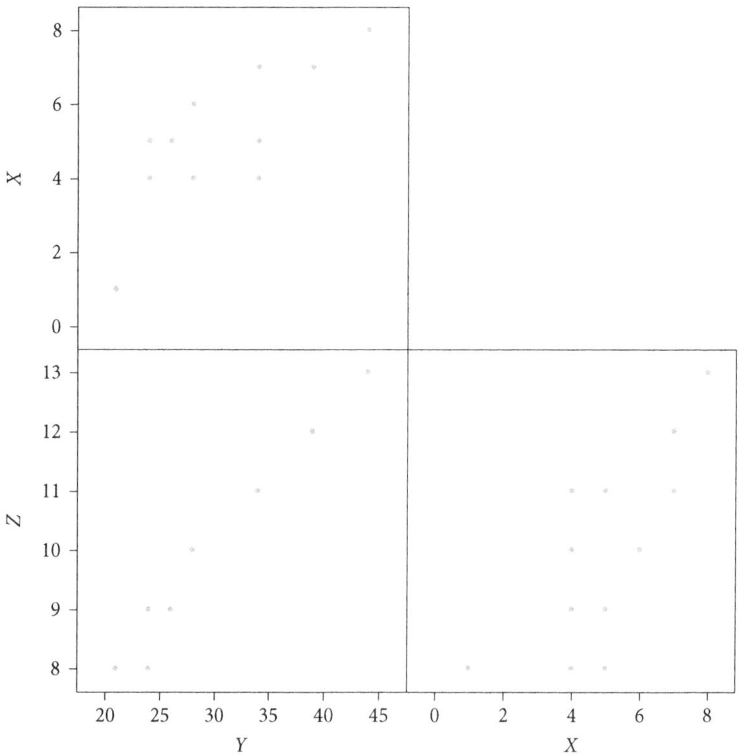

FIGURE 5.4 *Scatterplots of various salary, education and experience pairs*

The correlation between consumption and income is 0.997, indicative of the tightness of the association between the two variables, but the correlation between output growth and unemployment is just −0.07, so that these two variables are effectively uncorrelated and there appears to be no association between them.

5.3 Outliers and rank correlation

Like many summary statistics, the correlation coefficient can be heavily influenced by outliers. Consider the scatterplot shown in Figure 5.5, in which the data is obviously 'contaminated' by a single outlying observation, otherwise there would be a perfect correlation of +1.

As it is, the sample correlation coefficient is only 0.816 – large, but some way below unity. One way of dealing with the outlier is to compute a (*Spearman's*) *rank correlation coefficient*. Rather than using the raw data, this uses the *ranks* of the data instead, these being the position of each value of a variable when the values are ranked in ascending order (as in the computation of the median in §2.2). The calculations that are

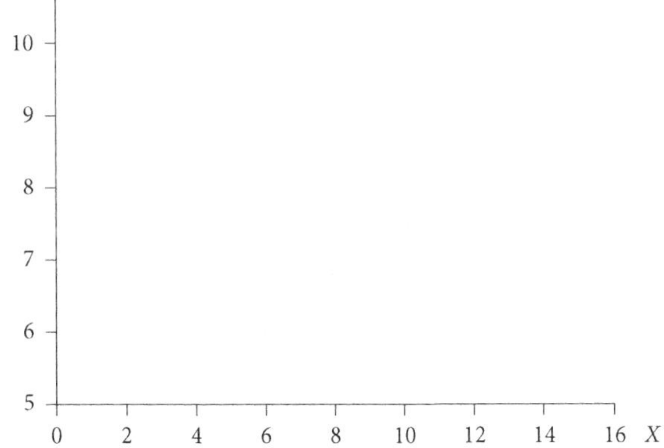

FIGURE 5.5 *Perfect correlation contaminated by an outlier*

TABLE 5.3 *Calculations for the rank correlation coefficient of the data in Figure 5.5*

X	Y	Rank(X)	Rank(Y)	d	d^2
4	5.39	1	1	0	0
5	5.73	2	2	0	0
6	6.08	3	3	0	0
7	6.42	4	4	0	0
8	6.77	5	5	0	0
9	7.11	6	6	0	0
10	7.46	7	7	0	0
11	7.81	8	8	0	0
12	8.15	9	9	0	0
13	12.74	10	11	-1	1
14	8.84	11	10	1	1

required to compute a rank correlation for the data in Figure 5.5 are shown in Table 5.3.

Here d = rank(X)–rank(Y) is the difference in ranks, and the rank correlation is given by

$$r_{XY}^{s} = 1 - \frac{6\sum d^2}{N\left(N^2 - 1\right)} = 1 - \frac{6 \times 2}{11\left(11^2 - 1\right)} = 0.991$$

Thus the rank correlation, at 0.991, is much closer to unity than is the standard correlation coefficient.

In the absence of outliers, the rank correlation will be similar to the ordinary correlation: for example, for the employee data set, $r_{XY}^{s} = 0.735$, $r_{ZY}^{s} = 0.980$ and $r_{XZ}^{s} = 0.680$.

5.4 Correlation and causation

It is often tempting to conclude that when a large correlation between two variables is found, one of the variables in some sense *causes* the other.[3] Such a temptation should be resisted because, unless we are able to invoke a feasible causal theory that posits that changes in one variable produces changes in the other, *correlation does not imply causation.* Sometimes there is a 'natural' causal ordering: it is hard to believe that the large correlation found between salary and years of post-school education reflects other than a causal link from the latter to the former, as the reverse link would be 'time inconsistent'. However, the negative correlation between inflation and unemployment found in Figure 5.1 could

TABLE 5.4 *Data on two non-linearly related variables: $Y^2 + X^2 = 9$*

	Y	X	$YX = yx$	$Y^2 = y^2$	$X^2 = x^2$
1	0	−3	0	0	9
2	2.236	−2	−4.472	5	4
3	2.828	−1	−2.828	8	1
4	3.000	0	0	9	0
5	2.828	1	2.828	8	1
6	2.236	2	4.472	5	4
7	3.000	3	0	0	9
8	-2.236	−2	4.472	5	4
9	-2.828	−1	2.828	8	1
10	-3.000	0	0	9	0
11	2.828	1	−2.828	8	1
12	2.236	2	−4.472	5	4
Σ	0	0	0	70	38

just as well be argued to represent a causal effect running from unemployment to inflation (high unemployment leads to lower demands for wage increases and hence lower inflation) as one running from inflation to unemployment (workers price themselves out of a job by demanding wages that keep up with inflation). Even the consumption–income relationship is by no means clear cut: the consumption function states that consumption is a function of income, but the national accounting identity has income defined as the sum of consumption, investment, government expenditure, etc., thus making the relationship one of *simultaneity*, that is, the two variables *jointly* influence each other.[4]

5.5 Further pitfalls in correlation analysis

Consider the data on the variables X and Y tabulated in Table 5.4.

The correlation between the two is clearly zero since the covariance is $\sum xy = 0$, and hence it appears that X and Y are unrelated. However, the scatterplot of the two variables, shown in Figure 5.6, belies that conclusion: it shows that they are, in fact, *perfectly* related, but that the relationship is *non-linear*, as all the data points lie on the circle $Y^2 + X^2 = 9$, so that $Y = \sqrt{9 - X^2}$.

This illustrates the important point that correlation is a measure of *linear association* and will not necessarily correctly measure the strength of a non-linear association. Thus, for example, the correlation of −0.34 between inflation and unemployment may well underestimate the

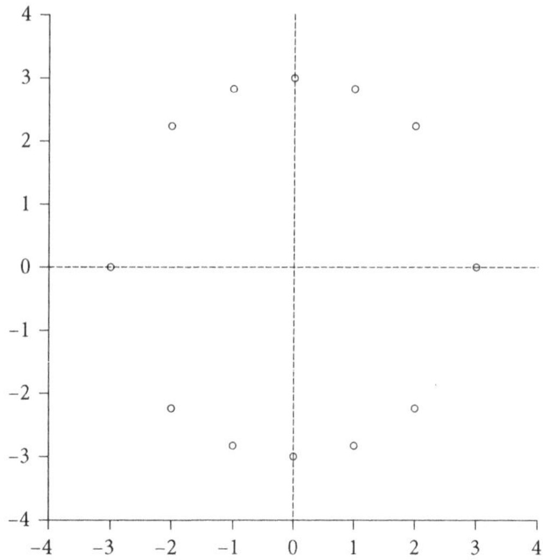

FIGURE 5.6 *Scatterplot of the data in Table 5.5 showing the relationship* $Y^2 + X^2 = 9$

strength of the relationship if, as suggested by the earlier Phillips curve analysis of §3.3, it is really a non-linear one.

Another way that correlation can give a misleading measure of association is when the observed correlation between two variables is a consequence of both being related to a third variable. This gives rise to the phenomenon known as *spurious correlation.*

Salary, education and experience revisited

In this example we found that all three variables were highly correlated (the correlation coefficients were calculated as 0.783, 0.976 and 0.740), so that it is possible that the large positive correlation between, for example, salary and education may be due to both variables being highly correlated with years of work experience.

To ascertain whether this may be the case, we can calculate the set of *partial correlation coefficients*. These coefficients measure the correlation between two variables with the influence of the third removed or, to be more precise, 'held constant'. The partial correlation between Y and X with Z held constant is defined as[5]

$$r_{XY.Z} = \frac{r_{XY} - r_{XZ}r_{YZ}}{\sqrt{1-r_{XZ}^2}\sqrt{1-r_{YZ}^2}}$$

Only if both r_{XZ} and r_{YZ} are both zero, that is, if both X and Y are uncorrelated with Z, will the partial correlation $r_{XY.Z}$ be identical to the 'simple' correlation r_{XY}. Thus, the partial correlation between salary and education with experience held constant is

$$r_{XY.Z} = \frac{0.783 - (0.740 \times 0.976)}{\sqrt{1-0.740^2}\sqrt{1-0.976^2}} = 0.414$$

This is rather smaller than the simple correlation, and suggests that the strong positive correlation between salary and education may be spurious and could be a 'statistical artefact' produced by omitting experience from the analysis. Similarly

$$r_{XZ \cdot Y} = \frac{r_{XZ} - r_{XY} r_{ZY}}{\sqrt{1-r_{XY}^2}\sqrt{1-r_{ZY}^2}} = -0.178$$

and

$$r_{YZ \cdot X} = \frac{r_{YZ} - r_{XY} r_{ZX}}{\sqrt{1-r_{XY}^2}\sqrt{1-r_{ZX}^2}} = 0.948$$

The latter partial correlation shows that the strength and sign of the association between salary and experience holds up on taking account of education, so that it is really experience that is the 'driver' of salary, rather than education.

Consumption, income and time trends

Figure 5.7 presents the time series plots of the consumption (C) and income (Y) series that produced the scatterplot in Figure 5.2.

Both have pronounced upward movements throughout the sample period, that is, they have *time trends*. This is behaviour typical of macroeconomic aggregates, but prompts the question of whether the very strong correlation between the two series (0.999) might, at least in part, be a consequence of a shared correlation with time itself. A time trend variable can be defined as $t = 1, 2, 3, \ldots$, i.e., it takes the value 1 for the first observation of the sample, here 1948, the value 2 for the second observation (1949), and so on up to 63 for the last observation, 2010.

The simple correlations between consumption and t and income and t are $r_{Ct} = 0.969$ and $r_{Yt} = 0.976$, respectively, confirming the presence of

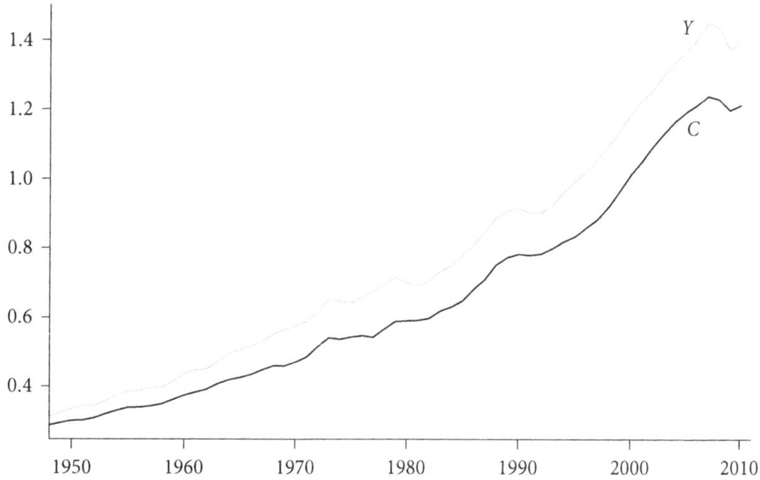

FIGURE 5.7 *Real consumption (C) and income (Y), 1948–2010*

the time trends in each variable. The partial correlation between consumption and income is then

$$r_{CY.t} = \frac{r_{CY} - r_{Ct}r_{Yt}}{\sqrt{1-r_{Ct}^2}\sqrt{1-r_{Yt}^2}}$$

$$= \frac{0.999 - (0.969 \times 0.976)}{\sqrt{1-0.969^2}\sqrt{1-0.976^2}} = 0.988$$

Although a little smaller than the simple correlation, the partial correlation is still large, so that the strong association between consumption and income is not spurious and is not simply a consequence of their sharing a common time trend. This, of course, is good news for macroeconomics as the relationship between the two variables is central to all models of the macroeconomy!

Notes

1 There are several ways of showing this inequality. One that requires nothing more than summations and basic algebra is to define $x_i = X_i - \bar{X}$ and $y_j = Y_j - \bar{Y}$ and to consider the double summation

$$\sum_i \sum_j \left(x_i y_j - x_j y_i\right)^2$$

which must clearly be non-negative. This expression may be expanded as

$$\sum_i x_i^2 \sum_j y_j^2 + \sum_j x_j^2 \sum_i y_i^2 - 2\sum_i x_i y_i \sum_j x_j y_j = 2\left(\sum_i x_i^2 \sum_i y_i^2 - \left(\sum_i x_i y_i\right)^2\right) \geq 0$$

so that it must be the case that $r_{XY}^2 = \left(\sum xy\right)^2 / \sum x^2 \sum y^2 \leq 1$ and hence that $-1 \leq r_{XY} \leq 1$. The expression for r_{XY}^2 is known as the *Cauchy–Schwarz inequality*. The square of the correlation coefficient, r_{XY}^2, will be shown to be a useful statistic for measuring the goodness of fit of regressions in **§6.14**.

2 It is perhaps easiest to show this equivalence by considering the numerator and denominator of (5.2) separately. Beginning with the numerator, we have

$$\begin{aligned}\sum\left(X - \bar{X}\right)\left(Y - \bar{Y}\right) &= \sum\left(XY - X\bar{Y} - \bar{X}Y + \bar{X}\bar{Y}\right) \\ &= \sum XY - \frac{\sum X \sum Y}{N} - \frac{\sum X \sum Y}{N} + \frac{\sum X \sum Y}{N} \\ &= N^{-1}\left(N\sum XY - \sum X \sum Y\right)\end{aligned}$$

which is N^{-1} times the numerator of (5.1). Similarly for the denominator of (5.2):

$$\begin{aligned}\sum\left(X - \bar{X}\right)^2 \sum\left(Y - \bar{Y}\right)^2 &= \sum\left(X^2 - 2X\bar{X} + \bar{X}^2\right)\left(Y^2 - 2Y\bar{Y} + \bar{Y}^2\right) \\ &= \left(\sum X^2 - 2\frac{(\sum X)^2}{N} + \frac{(\sum X)^2}{N}\right)\left(\sum Y^2 - 2\frac{(\sum Y)^2}{N} + \frac{(\sum Y)^2}{N}\right) \\ &= \left(\sum X^2 - \frac{(\sum X)^2}{N}\right)\left(\sum Y^2 - \frac{(\sum Y)^2}{N}\right) \\ &= N^{-2}\left(N\sum X^2 - (\sum X)^2\right)\left(N\sum Y^2 - (\sum Y)^2\right)\end{aligned}$$

which, on taking the square root, gives N^{-1} times the denominator of (5.1). On taking the ratio of these two expressions, the factor N^{-1} cancels out, thus showing the equivalence of (5.1) and (5.2).

3 We do not attempt to give a formal definition of what is meant by causality, as this is still the subject of great philosophical debate! Two important, but technically difficult, references are Arnold Zellner, 'Causality and econometrics', in K. Brunner and A.H. Meltzer (editors), *Three Aspects of Policy and Policymaking: Knowledge, Data and Institutions*, Carnegie–Rochester Conference Series on Public Policy 10 (1979), 9–54, and Paul W. Holland, 'Statistics and causal inference', *Journal of the American Statistical Association*

81 (1986), 945–960. A recent and more accessible discussion is Kevin Hoover, 'Causality in economics and econometrics', in S.N. Durlauf and L.F. Blume (editors), *The New Palgrave Dictionary of Economics, 2nd edition* (Palgrave Macmillan, 2008). The concept of *Granger causality* is widely used in economics: see, for example, Clive W.J. Granger, 'Testing for causality: a personal viewpoint', *Journal of Economic Dynamics and Control* 2 (1980), 329–352.

4 A simple national accounting identity has $Y = C + I + G$, to use a standard notation, so that if the consumption function has the linear form $C = a + bY$ then it is clear that Y and C are simultaneously determined. Discussions of this fundamental problem in applied economics may be found in the chapters by Aris Spanos, Kevin Hoover, and Duo Qin and Christopher Gilbert in Terence C. Mills and Kerry Patterson (editors), *The Palgrave Handbook of Econometrics, Volume 1: Econometric Theory* (Palgrave Macmillan, 2006). This problem is considered further in §16.2.

5 A derivation of this formula is given in §6.4.

6
Regression

Abstract: *Regressions are introduced as straight lines fitted through a scatterplot. The calculation of a regression as the 'line of best fit', obtained by minimising the sum of squared vertical deviations about the line (the least squares approach), is developed. This provides the least squares formulae for estimating the intercept and slope, and the interpretation of the regression line is discussed. The links between correlation, causation, reverse regression and partial correlation are investigated. Further issues involving regressions, such as how to deal with non-linearity, the use of time trends and lagged dependent variables as regressors, and the computation of elasticities, are all developed and illustrated using various economic examples.*

6.1 Correlation and regression

In the discussion of correlation in Chapter 5, a fitted line was often superimposed on scatterplots to help visually assess the strength of the association between the two variables (recall Figures 5.1–5.3). These fitted lines are in fact *regressions* and, while closely related to correlation coefficients, they offer a more sophisticated and powerful way of examining the relationship between two (or indeed more) variables. These strengths will be brought out in detail during this chapter, but we begin by focusing on how a regression line is calculated.

6.2 Calculating a regression: the 'line of best fit'

The statistical set-up of regression is the same as that for correlation: a sample of N pairs of observations on the variables X and Y

$$(X_1, Y_1), (X_2, Y_2), \ldots, (X_N, Y_N)$$

Rather than compute the correlation between the two variables, we now wish to fit a (sample) regression line. Suppose we choose to fit the line

$$\hat{Y}_i = a + bX_i \tag{6.1}$$

Here $\hat{Y}_i$ is the *predicted* value of Y *given* X_i, while the *intercept*, a, and *slope*, b, are coefficients. This is illustrated using the data on salary (Y) and post-school education (X) used in previous examples in Chapter 5 (see Tables 5.1 and 5.2). The scatterplot and fitted line are shown as Figure 6.1: the line is actually

$$\hat{Y} = 15.263 + 2.947X$$

How did we arrive at these numbers (known as *estimates*) of a and b? To find out, we first have to introduce the concept of a *regression residual*. This is the difference between the observed value Y_i and the value predicted by the regression line, $\hat{Y}_i$, that is, $e_i = Y_i - \hat{Y}_i$. Geometrically, it is the *vertical* distance from a point in the scatter to the fitted line.

Equivalently, we can write

$$Y_i = \hat{Y}_i + e_i \tag{6.2}$$

Substituting (6.1) into (6.2) gives

$$Y_i = a + bX_i + e_i \tag{6.3}$$

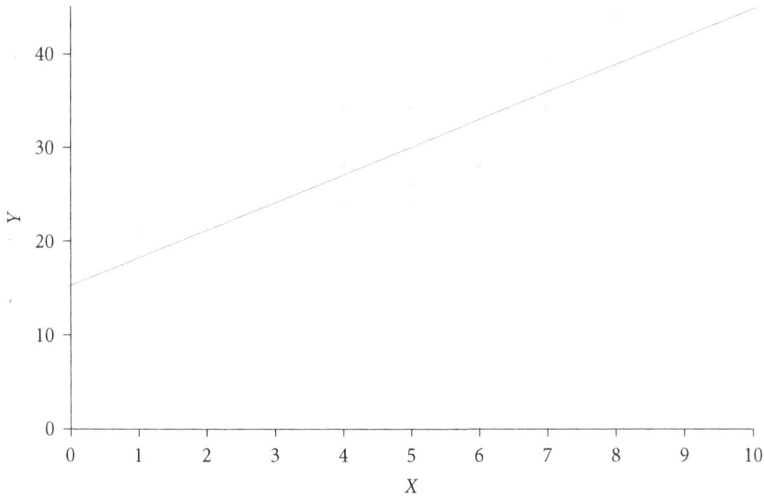

FIGURE 6.1 *Scatterplot of salary (Y) on education (X) with fitted line superimposed*

so that the regression line splits Y_i into two components: (i) a part 'explained' by X_i, given by $a + bX_i$; and (ii) the residual, e_i, that is left unexplained by X_i. In a good fit, part (i) should be as large as possible or, equivalently, the residuals should be as small as possible.

But what do we mean by the phrase 'the residuals should be as small as possible'? We actually mean that we should minimise some function of the residuals that will produce the 'line of best fit'. This function can be shown to be the 'sum of squared residuals', so that a and b are chosen to be those values which *minimise the sum of squared residuals* (geometrically, the sum of the squared *vertical* distances from the regression line): this is known as the *least squares criteria*.[1] Mathematically, the task is to choose those values of a and b that minimise

$$\sum_{i=1}^{N} e_i^2 = \sum_{i=1}^{N} \left(Y_i - a - bX_i\right)^2 \tag{6.4}$$

Finding a solution to (6.4) requires the use of differential calculus, and such a solution is provided in §12.1. The resulting formulae for the *least squares estimates* a and b are

$$b = \frac{N\sum XY - \sum X \sum Y}{N\sum X^2 - \left(\sum X\right)^2} = \frac{\sum xy}{\sum x^2} \tag{6.5}$$

$$a = \bar{Y} - b\bar{X} \tag{6.6}$$

For the salary and education data, we thus have

$$b = \frac{(12 \times 1912) - (60 \times 360)}{(12 \times 338) - 60^2} = \frac{112}{38} = 2.947$$

and

$$a = 30 - (2.947 \times 12) = 15.263.$$

6.3 Interpreting the regression line

The fitted regression line $\hat{Y} = 15.263 + 2.947X$ may be interpreted in the following way.

The intercept $a = 15.263$ is the value taken by $\hat{Y}$ when $X = 0$, i.e., it measures the predicted salary, £15,263, of an employee with no post-school education. The slope $b = 2.947$ measures the increment to salary for each additional year of post-school education: one year of such education is predicted to increase salary by 2.947 (that is, £2,947), two years is predicted to increase salary by $2.947 \times 2 = 5.894$, etc.

Presumably, this logic cannot be taken to extremes. The maximum X value in the sample is 8, that is, eight years of post-school education, which in itself seems to be taking the idea of a 'perpetual student' a bit far! Larger values of X are thus extremely unlikely, and even if they did occur might not follow the same linear relationship with Y, as the employer may not wish to pay such a large salary to someone who may be academically overqualified and/or dilatory. Care must therefore be taken when using regression fits to predict the behaviour of Y for values of X outside their usual range: this is known as the 'perils of extrapolation'.[2]

6.4 Correlation, causation and reverse regression

When discussing correlation, it was emphasised in §5.4 that unless we are able to invoke a feasible causal theory positing that a change in one variable produces a change in the other, correlation does not imply causation. In contrast, the set-up of a regression model implicitly assumes that such a causal theory exists by designating one variable, Y, to be the *dependent* (or *endogenous*) variable and the other, X, to be *independent* (or *exogenous*): the equation $Y = a + bX$ formalises the theory that changes in X produce changes in Y and not vice versa. Regression thus goes a stage further than correlation by assuming a causal link running from X to Y.

As discussed earlier, salary and years of post-school education have a natural, temporal causal ordering running from education to salary, which was why we estimated the regression model above. As was pointed out, however, in other cases such a causal ordering may be less clear-cut, so that regressing X on Y may be just as legitimate as regressing Y on X. Given that we are following the usual convention of plotting the Y-variable on the vertical axis and the X-variable on the horizontal, the *reverse regression* of X on Y is, geometrically, obtained by minimising the sum of squared *horizontal distances* from the regression line. Mathematically, the formulae for estimating the intercept c and slope d of the line $\hat{X}_i = c + dY_i$ are obtained by interchanging X and Y in equations (6.5) and (6.6):

$$d = \frac{N\sum XY - \sum X \sum Y}{N\sum Y^2 - \left(\sum Y\right)^2} = \frac{\sum xy}{\sum y^2}$$

$$c = \bar{X} - d\bar{Y}$$

Thus, for the regression of post-school education on salary, *even though the regression is clearly theoretically inappropriate*, we obtain

$$d = \frac{112}{538} = 0.208 \qquad\qquad c = 5 - (0.208 \times 30) = -1.245$$

An interesting result concerns the product of the slopes of the direct (Y on X) and reverse (X on Y) regressions, which is

$$bd = \frac{\left(\sum xy\right)^2}{\sum x^2 \sum y^2} = r_{XY}^2$$

that is, the product of the slope coefficients is the *squared correlation* between X and Y. If r_{XY}^2 is close to 1, $b \approx 1/db$, and the two regressions will be close to each other. Note also that

$$r_{XY}^2 = \frac{\sum xy}{\sum x^2}\frac{\sum xy}{\sum y^2} = b\frac{s_{XY}}{s_Y^2} = \frac{\sum xy}{\sum y^2}\frac{\sum xy}{\sum x^2} = d\frac{s_{XY}}{s_X^2}$$

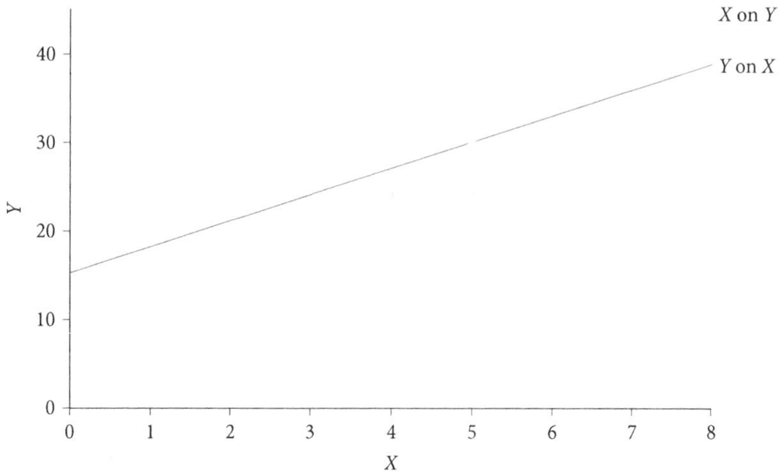

FIGURE 6.2 *Direct and reverse regressions of Y and X*

Here, we have

$$bd = 2.947 \times 0.208 = 0.614 = 0.783^2$$

so that the direct and reverse regressions are distinct, as can be seen from Figure 6.2.

The quantity r_{XY}^2 has an important interpretation: it measures the proportion of the variation in the dependent variable that is explained by the regression. To show this, substitute (6.6) into (6.4):

$$\begin{aligned}\sum e^2 &= \sum\left(Y - \bar{Y} + b\bar{X} - bX\right)^2 = \sum\left(y - bx\right)^2 = \sum\left(y^2 - 2bxy + b^2x^2\right)\\ &= \sum y^2 - 2b\sum xy + b^2\sum x^2\\ &= \sum y^2 - 2\frac{\left(\sum xy\right)^2}{\sum x^2} + \frac{\left(\sum xy\right)^2}{\sum x^2}\\ &= \sum y^2 - \frac{\left(\sum xy\right)^2}{\sum x^2}\\ &= \sum y^2\left(1 - \frac{\left(\sum xy\right)^2}{\sum x^2 \sum y^2}\right)\\ &= \sum y^2\left(1 - r_{XY}^2\right)\end{aligned} \tag{6.7}$$

Thus

$$r_{XY}^2 = 1 - \frac{\sum e^2}{\sum y^2} = \frac{\sum y^2 - \sum e^2}{\sum y^2}$$

Since $\sum y^2 - \sum e^2$ is clearly that part of the variation in y that is explained by the regression, r_{XY}^2 is the proportion of that variation explained by the regression. Thus $r_{XY}^2 = 0.614$ implies that 61.4% of the variation in Y is explained by the regression $\hat{Y} = 15.263 + 2.947X$, leaving 38.6% unexplained and captured by the residual. For this reason r_{XY}^2 is often referred to as a measure of *goodness of fit*. Note that $r_{XY}^2 = 0.614$ also implies that 61.4% of the variation in X is explained by the reverse regression $\hat{X} = -1.245 + 0.208Y$.

Of course, a positive correlation implies a positive relationship between the two variables, which in turn implies that the slope b will be positive. This can be shown formally through

$$r_{XY} = \frac{\sum xy}{\sqrt{\sum x^2}\sqrt{\sum y^2}} = \frac{\sum xy}{\sum x^2}\frac{\sqrt{\sum x^2/(N-1)}}{\sqrt{\sum y^2/(N-1)}} = b\frac{s_X}{s_Y} \qquad (6.8)$$

Since the ratio of the sample standard deviations s_X/s_Y must be positive, r_{XY} and b must be of the same sign (or both be zero).

A link with partial correlations

The result in (6.7) may be used to derive the partial correlation coefficient introduced in §5.5. Consider the residuals from the regressions of Y on X, Y on Z and X on Z, using an obvious distinguishing notation:

$$e_{YX} = y - b_{YX}x, \qquad e_{YZ} = y - b_{YZ}z \qquad e_{XZ} = x - b_{XZ}z$$

where we use the result implicit in the first line of (6.7) that equation (6.3), $Y = a + bX + e$, and $y = bx + e$ are equivalent ways of expressing a regression depending upon whether the data is measured in raw or mean deviation form. The partial correlation $r_{YX.Z}$ is then defined as the correlation between e_{YZ} and e_{XZ}, the residuals from the Y on Z and X on Z regressions, respectively, that is, it is the correlation between Y and X with the common influence of Z having first been 'purged' by regressing Y and X on Z:[3]

$$r_{YX.Z} = \frac{\sum e_{YZ} e_{XZ}}{\sqrt{\sum e_{YZ}^2 \sum e_{XZ}^2}}$$

From (6.7), and using the expressions in §5.2, we have

$$\sum e_{YZ}^2 = (N-1) s_Y^2 \left(1 - r_{YZ}^2\right)$$

$$\sum e_{XZ}^2 = (N-1) s_X^2 \left(1 - r_{XZ}^2\right)$$

Some algebra shows that the covariance between e_{YZ} and e_{XZ} can be written as[4]

$$\sum e_{YZ} e_{XZ} = (N-1) s_X s_Y \left(r_{XY} - r_{YZ} r_{XZ}\right)$$

so that it follows immediately that

$$r_{YX.Z} = \frac{\sum e_{YZ} e_{XZ}}{\sqrt{\sum e_{YZ}^2 \sum e_{XZ}^2}} = \frac{r_{XY} - r_{YZ} r_{XZ}}{\sqrt{\left(1 - r_{YZ}^2\right)\left(1 - r_{XZ}^2\right)}}$$

6.5 Dealing with non-linearity

In discussing transformations of variables in §3.3, we presented a fitted Phillips curve, namely

$$\pi = 1.4 + 4.4 \frac{1}{U} \tag{6.9}$$

This non-linear function can be estimated by our (linear) regression technique by defining $Y = \pi$ (inflation) and $X = 1/U$ (inverse unemployment). As long as the function is linear in the *coefficients* a and b, as it is here ($\pi = a + bU^{-1}$), then linear regression can be used after the transformed variables have been computed. Thus, for example, *powers*, such as X^2 and X^3, are straightforward to incorporate, as are logarithms and many other transformations.

Interestingly, the fitted Phillips curve (6.9) may be compared to the linear regression fit, which may be calculated to be

$$\pi = 6.0 - 0.6U$$

The two fits are shown graphically in Figure 6.3, but it is difficult to assess from this figure which of the two equations should be preferred.

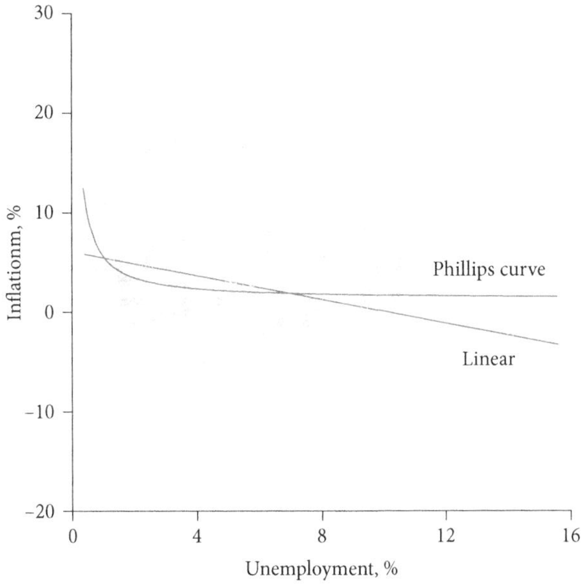

FIGURE 6.3 *Phillips curve and linear fits to UK inflation and unemployment rates, 1855–2011*

The goodness of fit measures from the two regressions are $r^2_{\pi,1/U} = 0.108$ and $r^2_{\pi U} = 0.113$ so that, in fact, the linear regression provides a slightly better fit than the Phillips curve, although both only explain ~11% of the variation in inflation. This suggests that there is only a very weak relationship between inflation and unemployment over the last 150 years or so in the U.K.

6.6 Regression on time trends and lagged dependent variables

In discussing spurious correlation in §5.5, we argued that both consumption and income in the UK contained time trends, as both series were very highly correlated with time. Given the link between correlation and regression, another way of analysing this feature is to consider using time as the independent variable (sometimes referred to as the *regressor*) in a regression, that is, if we have T observations on the variable Y, $Y_1, Y_2, \ldots, Y_T$, then the *time trend regression* is

$$Y_t = a + bt + e_t$$

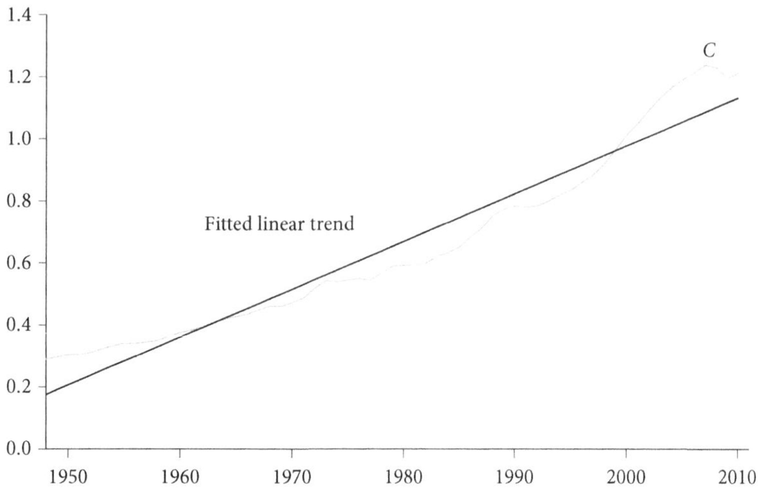

FIGURE 6.4 *U.K. consumption, 1948–2010, with fitted trend* $\hat{C} = 158600 + 15460t$

Figure 6.4 plots the U.K. consumption series (C) and superimposes upon it the fitted linear trend $\hat{C} = 158600 + 15460t$, which has a goodness of fit of $r_{Ct}^2 = 0.939 = 0.969^2$.

This trend has the interpretation that as t increases by one (year) C increases by 15460; in other words, as each year passes the trend in consumption increases by this amount, starting from $158600 + 15460 = 174060$ in year 1 (1948).

Although the goodness of fit is pretty high at 94%, the linear trend fails to capture the modest curvature in the series, as the trend under-predicts consumption in the early and later years of the sample and over-predicts it during the central years. This non-linearity could be captured by fitting a *quadratic trend*, which would require the technique of multiple regression (to be discussed in Chapter 13), but a better alternative here is to follow the approach of §3.2, and first linearise consumption by taking logarithms and then fit a linear trend. If we do this we obtain $r_{\ln C,t}^2 = 0.993$ and

$$\ln\hat{C} = 12.52 + 0.024t$$

The plot of $\ln C$ with its fitted trend is shown in Figure 6.5, and this provides a superior fit to the linear trend of Figure 6.4 in the sense that the

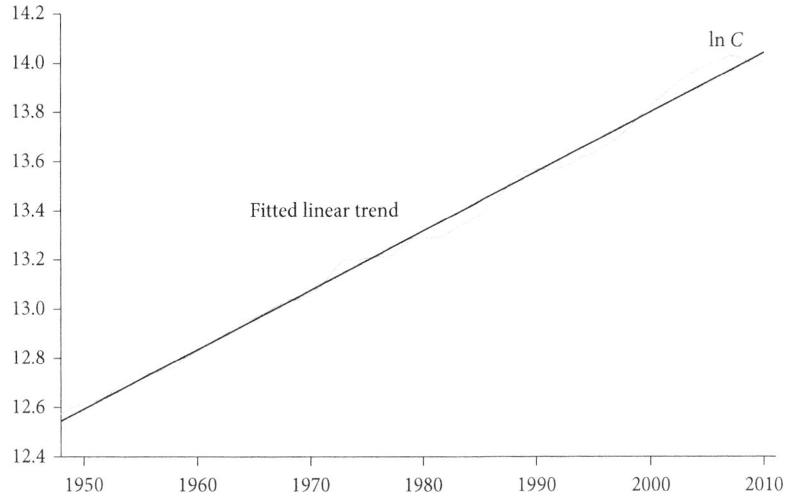

FIGURE 6.5 *Logarithms of UK consumption with fitted linear trend*
$\ln\hat{C} = 12.52 + 0.024t$

linear evolution of the logarithm of consumption is now closely fitted by the time trend.

The regression of $\ln Y$ on t has, in fact, a very important interpretation. At time t, the trend model is given by

$$\ln\hat{Y}_t = a + bt \tag{6.10}$$

whereas one time period earlier, at t–1, it is given by

$$\ln\hat{Y}_{t-1} = a + b(t-1) \tag{6.11}$$

Subtracting (6.11) from (6.10) gives

$$\ln\hat{Y}_t - \ln\hat{Y}_{t-1} = a - a + bt - bt + b$$

which is, on recalling §3.2,

$$\ln\hat{Y}_t - \ln\hat{Y}_{t-1} = \Delta\ln\hat{Y}_t = b$$

In other words, the slope coefficient b can be interpreted as the *trend growth rate* of Y. For consumption, $b = 0.024$, so that trend consumption growth over the period 1948–2010 has been 2.4% per annum.

In models of this type, the residuals $e_t = \ln Y_t - a - b_t$ also have an interesting interpretation as the *deviation from the trend growth path of Y*. For macroeconomic aggregates like consumption, income and output, such deviations can be thought of as *business cycle fluctuations*. Thus, if we look carefully at Figure 6.5, negative residuals (when the actual value is below trend) correspond to periods of recession, while positive residuals correspond to periods of above-trend growth, expansion. For example, the decade from the late 1970s to the late 1980s saw consumption below trend, which was thus a recessionary period, while the late 1980s to the early 1990s and the years from 1997 saw above-trend growth until 2009, when the recessionary impact of the credit crunch began to be felt.

Fitting a linear trend to the logarithms of income gives $\ln \hat{Y}_t = 12.66 + 0.025t$, which is very similar to the consumption trend. One of the key theories of aggregate consumption behaviour is that over long periods of time trend growth in consumption and income should be the same, and this appears to be borne out by the UK data.[5]

An alternative way of dealing with trending behaviour is to explain the current level of a variable, X_t, by the value it took last period, X_{t-1}; we say that we include the *lagged dependent variable* as a regressor:

$$X_t = a + bX_{t-1} + e_t$$

If we have data from $t = 1$ up to T, then a regression can only be run on the last $T-1$ observations, that is, from $t = 2$ to T, because for $t = 1$, $X_{t-1} = X_0$, which is the value of X immediately before the sample and is therefore not available to us. Thus the formula for the slope coefficient estimate is

$$b = \frac{\sum_{t=2}^{T} x_t x_{t-1}}{\sum_{t=1}^{T-1} x_t^2}$$

For the logarithm of income, the lagged dependent variable regression is

$$\ln \hat{Y}_t = 0.113 + 0.993 \ln \hat{Y}_{t-1}$$

The estimate of b is seen to be very close to 1: if $b = 1$ then we would have the general model

$$X_t = a + X_{t-1} + e_t$$

This is known as a *random walk (with drift)* and is a very important model in theories of dynamic economic behaviour, particularly financial theories.[6] Not every economic series is a random walk or necessarily close to being one. The lagged dependent variable model for inflation over the same period beginning in 1948 is

$$\pi_t = 1.23 + 0.78\pi_{t-1} + e_t$$

Here $b = 0.78$, which is some way below 1. The size of b in models of this type can be thought of as measuring the degree of *persistence* in a time series, that is, how strongly the current value is related to past values. For the UK, it would thus appear that income is much more persistent than inflation, but this would be a flawed comparison, for we are not comparing like with like. Recall that inflation is the growth rate of the price level, so for a fair comparison we should compare the lagged dependent variable model for inflation with a similar model for income growth, $\Delta \ln Y_t$. Regressing income growth on its lagged value gives $b = 0.34$, so that inflation is, in fact, rather more persistent than income growth.

6.7 Elasticities

A very important concept in economics is that of *elasticity*, which is the *proportionate* change in Y in response to a *proportionate* change in X. Mathematically, the elasticity is defined as

$$\eta_{YX} = \frac{dY/Y}{dX/X} = \frac{dY}{dX}\frac{X}{Y}$$

If we have the linear relationship $Y = a + bX$ then, since $dY/dX = b$, the elasticity is

$$\eta_{YX} - b\frac{X}{Y}$$

If we have a fitted regression line, then the slope estimate can be used for b, but note that the elasticity depends on the pair of X and Y values and will thus change over the sample. As an example, the linear consumption function for the scatterplot in Figure 5.2 is estimated to be

$$\hat{C}_t = -4708 + 0.86Y_t$$

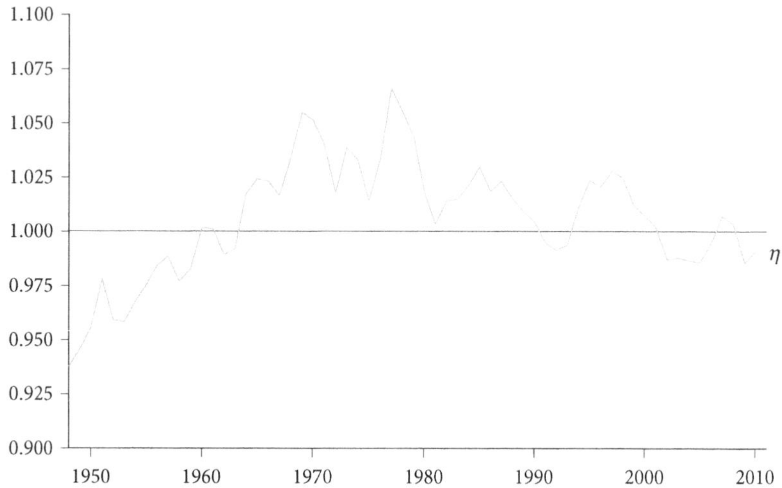

FIGURE 6.6 *Income elasticities of consumption, 1948–2010*

so that the *income elasticity of consumption* is[7]

$$\eta_t = 0.86\frac{Y_t}{C_t}$$

and the estimated elasticities for, say, 1948, 1980 and 2010 (at the beginning, in the middle and at the end of the sample period) are

$$\eta_{1948} = 0.86\frac{Y_{1948}}{C_{1948}} = 0.86\frac{314496}{288538} = 0.94$$

$$\eta_{1980} = 0.86\frac{Y_{1980}}{C_{1980}} = 0.86\frac{698528}{589955} = 1.02$$

$$\eta_{2010} = 0.86\frac{Y_{2010}}{C_{2010}} = 0.86\frac{1395312}{1211287} = 0.99$$

The complete series of elasticities is shown in Figure 6.6, and they are seen to fluctuate around unity throughout the sample period.

Rather than calculate the entire set, the *average elasticity* is often reported

$$\bar{\eta} = b\frac{\bar{Y}}{\bar{C}}$$

For the above data, we have

$$\bar{\eta} = 0.86\frac{764887}{653330} = 1.01$$

Thus the average income elasticity of consumption over the period 1948 to 2010 is very close to one, implying that, on average, an *x*% change in income produces an *x*% change in consumption.[8]

Suppose now that we have a *log-linear* consumption function

$$\ln C = a + b\ln Y$$

Standard results from calculus give

$$\frac{d\ln C}{d\ln Y} = b \qquad \frac{d\ln C}{dC} = \frac{1}{C} \qquad \frac{d\ln Y}{dY} = \frac{1}{Y}$$

so that

$$\frac{dC}{dY} = \frac{C}{Y}\frac{d\ln C}{d\ln Y} = \frac{C}{Y}b$$

and

$$\eta_{CY} = \frac{C}{Y}b\frac{Y}{C} = b$$

that is, for a log-linear model the elasticity is constant and given by the slope *b*. The estimated log-linear consumption function is

$$\ln\hat{C} = 0.09 + 0.98\ln Y$$

so that the elasticity is now estimated to be constant at 0.98, close to the unit elasticity result arrived at above.

Notes

1 The method of least squares is traditionally thought to have been originally proposed by Carl Friedrich Gauss in the early years of the 19th century, although there is now some dispute about this: see, for example, Robin L.

Plackett, 'The discovery of the method of least squares', *Biometrika* 59 (1972), 239–251, and Stephen M. Stigler, 'Gauss and the invention of least squares', *Annals of Statistics* 9 (1981), 463–474. Notwithstanding these debates over priority, least squares has since become one of the most commonly used techniques in statistics, and is the centrepiece of econometrics.

2 Mark Twain, *Life on the Mississippi* (Harper, 1883), chapter 22, gives an entertaining example of the perils of extrapolation:

> In the space of one hundred and seventy six years the Lower Mississippi has shortened itself two hundred and forty-two miles. That is an average of a trifle over a mile and a third per year. Therefore, any calm person, who is not blind or idiotic, can see that in the Old Oölitic Silurian Period, just a million years ago next November, the Lower Mississippi was upwards of one million three hundred thousand miles long, and stuck out over the Gulf of Mexico like a fishing-pole. And by the same token any person can see that seven hundred and forty-two years from now the Lower Mississippi will be only a mile and three-quarters long, and Cairo [Illinois] and New Orleans will have joined their streets together and be plodding comfortably along under a single mayor and a mutual board of aldermen. There is something fascinating about science. One gets such wholesale returns of conjecture out of such a trifling investment of fact.

3 Note that the residuals from any regression must have zero mean, since

$$\sum e=\sum(y-bx)=\sum y-b\sum x=0$$

as $\sum y=\sum x=0$ by definition.

4 Using (6.8) we have $b_{YZ}=r_{YZ}\left(s_Y/s_Z\right)$ and $b_{XZ}=r_{XZ}\left(s_X/s_Z\right)$, so that

$$\sum e_{YZ}e_{XZ}=\sum\left(y-r_{YZ}\left(s_Y/s_Z\right)z\right)\left(x-r_{XZ}\left(s_X/s_Z\right)z\right)$$

$$=\sum yx+r_{YZ}r_{XZ}\frac{s_Ys_X}{s_Z^2}\sum z^2-r_{YZ}\frac{s_Ys_Z}{s_Z^2}\sum zx-r_{XZ}\frac{s_Xs_Z}{s_Z^2}\sum zy$$

Substituting the following expressions

$$\sum yx=(N-1)r_{YX}s_Xs_Y,\quad \sum z^2=(N-1)s_Z^2,\quad \sum zx=(N-1)r_{ZX}s_Zs_X$$

$$\sum zy=(N-1)r_{ZY}s_Zs_Y$$

and simplifying then leads to the required expression.

5 The idea that consumption and income grow at the same trend rate is an implication of the *fundamental equation of neoclassical economic growth*, which has a steady-state solution which exhibits 'balanced growth'. The model is often referred to as the *Solow growth model*: Robert M. Solow, 'A contribution to the theory of economic growth', *Quarterly Journal of Economics* 71 (1956), 65–94.

6 The term *random* (or drunkard's) *walk* was first coined in a correspondence between Karl Pearson and Lord Rayleigh in the journal *Nature* in 1905. Although first employed by Pearson to describe a mosquito infestation in a forest, the model was subsequently, and memorably, used to describe the optimal search strategy for finding a drunk who had been left in the middle of a field at the dead of night! The solution is to start exactly where the drunk had been placed, as that point is an unbiased estimate (see **§11.1**) of the drunk's future position since he will presumably stagger along in an unpredictable and random fashion: '(t)he lesson of Lord Rayleigh's solution is that in open country the most probable place to find a drunken man who is at all capable of keeping on his feet is somewhere near his starting point'. In fact, random walks were first formally introduced in 1900 by Louis Bachelier in his doctoral dissertation *Théorie de Speculation*, although he never used the term. Under the supervision of the famous mathematician and polymath Henri Poincaré, Bachelier developed the mathematical framework of random walks in continuous time (where it is termed *Brownian motion*) in order to describe the unpredictable evolution of stock prices. The dissertation remained unknown until it was rediscovered in the mid-1950s after the mathematical statistician Jimmie Savage had come across a later book by Bachelier on speculation and investment. A translation of the dissertation ('Theory of speculation') by James Boness was eventually published in 1964 in Paul Cootner (editor), *The Random Character of Stock Market Prices* (MIT Press), pp. 17–78. Random walks were independently discovered by Albert Einstein in 1905 and, of course, have since played a fundamental role in mathematics and physics as models of, for example, waiting times, limiting diffusion processes, and first-passage-time problems. A popular account of the model and its implications for many facets of life is Leonard Mlodinow, *The Drunkard's Walk: How Randomness Rules Our Lives* (Penguin, 2009).

7 Note that we have dropped the YX subscripts on η to ease 'notational clutter' but have added t subscripts to emphasise that the elasticity changes over time. Also, be aware of potential confusion caused by income, the independent variable, being denoted Y_t.

8 This unit elasticity result is a key component of the popular *error correction model* of consumption; see, for example, James Davidson, David Hendry, Frank Srba and Stephen Yeo, 'Econometric modelling of the aggregate time series relationship between consumer's expenditure and income in the United Kingdom', *Economic Journal* 88 (1978), 661–692.

7

Basic Concepts of Probability

Abstract: *After discussing some introductory concepts in statistical inference and contrasting these with our previous emphasis on exploratory data analysis, the basic axioms of probability are introduced, along with the additive and multiplication rules for computing probabilities of compound events. To aid in such computations, the counting rules for combinations and permutations are introduced. Bayes theorem is discussed, as are the various definitions of probability – classical, relative frequency and subjective – and it is emphasised that all definitions follow the same axioms and rules.*

7.1 Probability theory and statistical inference

Our analysis of data has so far been purely *descriptive*, in the sense that we have taken the data 'as it comes' and have attempted to investigate its basic features: this is often known as *exploratory data analysis*.[1] While this is an essential first step in analysing any set of data, it does have its limitations and is unable to provide answers to many interesting and important questions that economists may wish to ask. This is primarily because no attention has yet been paid as to how the observations – our *sample* of data – have been obtained. This might, at first sight, seem a curious question: after all, we typically just analyse the observations that are available to us. But some reflection on the data used in the examples of previous chapters should suggest that there are some intriguing differences in the way the observations have 'come to us'.

Consider first the observations on employee's salary, post-school education and years of work experience used in several examples. Now, it might be the case that this was a very small firm with just these 12 employees, so that the data consists of a *complete enumeration* of the *population* of the firm's employees. However, suppose that the firm has many more than 12 employees, so that our data represents a *sample from the population*. How can we be sure that this is a *representative* sample, one that reproduces the key features of the population? For example, are the sample means likely to be good estimates of the population mean salary, etc? From what we know, which is basically nothing, we cannot answer this question – the firm's senior management, who presumably have very high salaries, probably several years of post-school education, and almost certainly many years of work experience, may not feature in the sample, thus potentially biasing the sample means considerably.

Now recall the examples in which income for a 'wide cross-section of countries in 2009' was analysed. If you look again at Table 2.1, data for 189 countries was actually available, but the number of countries in the world is commonly thought to be larger than this, so that the analysis has again been carried out on an (admittedly large) sample from the population of countries.[2] Is this sample likely to be representative of the population as a whole? The missing data typically comes from countries whose income is likely to be relatively low. As a consequence, the mean and median income estimated from the available sample are almost certainly too high, and the histogram shown in Figure 2.1 probably underestimates the degree of skewness in the income distribution.

What about time series data? Surely this is a complete enumeration of the population, as the observations represent what has actually occurred in the past. Again, some reflection suggests potential difficulties. Table 2.2 lists annual inflation for the UK but, as we have seen, data on the underlying price index might be available, say, monthly. How was annual inflation calculated? Does it measure the proportionate rate of increase in prices from December of one year to December of the next, or does it measure it from June to June, or does it take the average price of the 12 months of one year and compare this with the corresponding average of the next year? All these will give different inflation values and hence, for example, different estimates of mean inflation. Is the estimate that we end up with representative of the population mean inflation rate?[3]

Whenever we have available just a sample of data from a population – and this, as we have seen, is the typical situation however the data is generated – then we need to be able to *infer* the (unknown) values of the population characteristics from their estimated sample counterparts. This is known as *statistical inference* and requires the knowledge of basic concepts in probability and an appreciation of probability distributions.

7.2 Basic concepts in probability

Many, if not all, readers will have had some exposure to the basic concepts of probability, so our treatment here will be accordingly brief. We will begin with a few definitions, which will enable us to establish a vocabulary.

- An *experiment* is an action, such as tossing a coin, which has a number of possible *outcomes* or *events*, such as heads (H) or tails (T).
- A *trial* is a single performance of the experiment, with a single outcome.
- The *sample space* consists of all possible outcomes of the experiment. The outcomes for a single toss of a coin are $[H, T]$, for example. The outcomes in the sample space are *mutually exclusive*, which means that the occurrence of one outcome rules out all the others: you cannot have both H and T in a single toss of the coin. They are also *exhaustive*, since they define all possibilities: you can only have either H or T in a single toss.

- With each outcome in the sample space we can associate a *probability*, which is the chance of that outcome occurring. Thus, if the coin is fair, the probability of H is 0.5.

With these definitions, we have the following *probability axioms.*

- The probability of an outcome A, $P(A)$, must lie between 0 and 1, i.e.

$$0 \leq P(A) \leq 1 \tag{7.1}$$

- The sum of the probabilities associated with all the outcomes in the sample space is 1. This follows from the fact that one, and only one, of the outcomes *must* occur, since they are mutually exclusive and exhaustive, that is, if $P(i)$ is the probability of outcome i occurring and there are n possible outcomes, then

$$\sum_{i=1}^{n} P(i) = 1 \tag{7.2}$$

- The *complement* of an outcome is defined as everything in the sample space apart from that outcome: the complement of H is T, for example. We will write the complement of A as $\bar{A}$ (to be read as 'not-A' and is not to be confused with the mean of A!). Thus, since A and $\bar{A}$ are mutually exclusive and exhaustive

$$P(\bar{A}) = 1 - P(A) \tag{7.3}$$

Most practical problems require the calculation of the probability of a set of outcomes, rather than just a single one, or the probability of a series of outcomes in separate trials; for example, what is the probability of throwing three H in five tosses? We refer to such sets of outcomes as *compound events.*

Although it is sometimes possible to calculate the probability of a compound event by examining the sample space, typically this is too complex, even impossible, to write down. To calculate compound probabilities in such situations, we make use of a few simple rules.

The addition rule

This rule is associated with 'or' (sometimes denoted as $\cup$, known as the *union* of two events):

$$P(A \text{ or } B) = P(A) + P(B) - P(A \text{ and } B) \tag{7.4}$$

Here we have introduced a further event, 'A and B', which encapsulates the idea of the *intersection* ($\cap$) of two events.

Consider the experiment of rolling a fair six-sided die.[4] The sample space is thus $[1,2,3,4,5,6]$, $n = 6$ and $P(i) = 1/6$ for all $i = 1,2,\ldots,6$. The probability of rolling a five or a six is

$$P(5 \text{ or } 6) = P(5) + P(6) - P(5 \text{ and } 6)$$

Now, $P(5 \text{ and } 6) = 0$, since a five and a six cannot simultaneously occur (the events are mutually exclusive). Thus

$$P(5 \text{ or } 6) = 1/6 + 1/6 - 0 = 1/3$$

It is not always the case that the two events are mutually exclusive, so that their intersection does not always have zero probability. Consider the experiment of drawing a single playing card from a standard $n = 52$ card pack. The sample space is the entire 52-card deck, and $P(i) = 1/52$ for all i. Let us now compute the probability of obtaining either a king or a heart from this single draw. Using obvious notation

$$P(K \text{ or } H) = P(K) + P(H) - P(K \text{ and } H) \tag{7.5}$$

Now, $P(K) = 4/52$ and $P(H) = 13/52$, obtained in each case by counting up the number of outcomes in the sample space that are defined by each event and dividing the result by the total number of outcomes in the sample space: for example, on extending the notation of (7.5) in an obvious way,

$$P(K) = P(K \text{ and } C) + P(K \text{ and } D) + P(K \text{ and } H) + P(K \text{ and } S)$$

However, by doing this, the outcome representing the king of hearts (K and H) gets included in both calculations and is thus 'double counted'. It must therefore be subtracted from $P(K) + P(H)$, thus leading to (7.5). The events K and H are *not* mutually exclusive, since $P(K \text{ and } H) = 1/52$, and

$$P(K \text{ or } H) = 4/52 + 13/52 - 1/52 = 16/52$$

The multiplication rule

The multiplication rule is associated with 'and'. Consider the event of rolling a die twice and asking what the probability of obtaining fives on both rolls is. Denote this probability as $P(5_1 \text{ and } 5_2)$, where we now use the notation i_j to signify obtaining outcome i on trial j: this probability will be given by

$$P(5_1 \text{ and } 5_2) = P(5_1) \times P(5_2) = \frac{1}{6} \times \frac{1}{6} = \frac{1}{36}$$

The logic of this calculation is straightforward. The probability of obtaining a five on a single roll of the die is 1/6 and, because the outcome of one roll of the die cannot affect the outcome of a second roll, the probability of this second roll also producing a five must again be 1/6. The probability of both rolls producing fives must then be the product of these two individual probabilities. Technically, we can multiply the individual probabilities together because the events are *independent.*

What happens when events are not independent? Consider drawing two playing cards and asking for the probability of obtaining a king of hearts (now denoted KH) on the first card and an H on the second. If the first card is replaced before the second card is drawn then the events KH_1 and H_2 are still independent, and the probability will be

$$P(KH_1 \text{ and } H_2) = P(KH_1) \times P(H_2) = \frac{1}{52} \times \frac{13}{52} = 0.00480$$

However, if the second card is drawn *without the first being replaced* (a 'normal' deal) then we have to take into account the fact that, if KH_1 has occurred, then the sample space for the second card has been reduced to 51 cards, of which only 12 will be H: thus the probability of H_2 occurring is 12/51, not 13/52. The probability of H_2 is therefore dependent on KH_1 having occurred, and thus the two events are not independent. Technically, this is a *conditional probability,* denoted here as $P(H_2|KH_1)$ and generally as $P(B|A)$; the general multiplication rule of probabilities is, for our example,

$$P(KH_1 \text{ and } H_2) = P(KH_1) \times P(H_2|KH_1) = \frac{1}{52} \times \frac{12}{51} = 0.00452$$

and, in general,

$$P(A \text{ and } B) = P(A) \times P(B|A)$$

Only if the events are independent will $P(B|A) = P(B)$. More formally, if A and B are independent, then, because the probability of B occurring is unaffected by whether A occurs or not,

$$P(B|A) = P(B|\bar{A}) = P(B)$$

and

$$P(A|B) = P(A|\bar{B}) = P(A)$$

Combining the rules

Consider the following example. Suppose that it is equally likely that a mother has a boy or a girl on the first (single) birth, but that a second birth is more likely to be a boy if there was a boy on the first birth, and similarly for a girl. Thus

$$P(B_1) = P(G_1) = 0.5$$

but, for the purposes of this example,

$$P(B_2|B_1) = P(G_2|G_1) = 0.6$$

which implies that

$$P(B_2|G_1) = P(G_2|B_1) = 0.4$$

The probability of a mother having one child of each sex is thus

$$\begin{aligned} P(1 \text{ girl and } 1 \text{ boy}) &= P((G_1 \text{ and } B_2) \text{ or } P(B_1 \text{ and } G_2)) \\ &= P(G_1)P(B_2|G_1) + P(B_1)P(G_2|B_1) \\ &= (0.5 \times 0.4) + (0.5 \times 0.4) \\ &= 0.4 \end{aligned}$$

Counting rules: combinations and permutations

The preceding problem can be illustrated using a *tree diagram*, which is a way of enumerating all possible outcomes in a sample space with their associated probabilities. Tree diagrams have their uses for simple problems, but for more complicated ones they quickly become very complex and difficult to work with.[5]

For more complicated problems, counting rules must be employed. Suppose we have a family of five children of whom three are girls. To compute the probability of this event occurring, our first task is to be able to calculate the number of ways of having three girls and two boys, irrespective of whether successive births are independent or not. An 'obvious' way of doing this is to write down all the possible orderings, of which there are ten:

GGGBB	GGBGB	GGBBG	GBGGB	GBGBG
GBBGG	BGGGB	BGGBG	BGBGG	BBGGG

In more complex problems, this soon becomes difficult or impossible, and we then have to resort to using the *combinatorial formula*. Suppose the three girls are 'named' *a*, *b* and *c*. Girl *a* could have been born first, second, third, fourth or fifth, that is, in any one of five 'places' in the ordering:

a???? *?a???* *??a??* *???a?* *????a*

Suppose that *a* is born first; then *b* can be born either second, third, fourth or fifth, that is, any one of four places in the ordering:

ab??? *a?b???* *a??b?* *a???b*

But *a* could have been born second, etc., so that the total number of places for *a* and *b* to 'choose' is 5×4 = 20. Three places remain for *c* to choose, so, by extending the argument, the three girls can choose a total of 5×4×3 = 60 places between them. This is the number of *permutations* of three *named* girls in five births, and can be given the notation

$$ {}_5P_3 = 5\times4\times3 = \frac{5\times4\times3\times2\times1}{2\times1} = \frac{5!}{2!} $$

which uses the *factorial* (!) notation.

${}_5P_3 = 60$ is six times as large as the number of possible orderings written down above. The reason for this is that the listing does not distinguish between the girls, denoting each of them as G rather than *a*, *b* or *c*. The permutation formula thus overestimates the number of combinations by a factor representing the number of ways of ordering the three girls, which is $3\times2\times1 = 3!$. Thus the formula for the *combination* of three girls in five births is

$$ {}_5C_3 = \frac{{}_5P_3}{3!} = \frac{5!}{3!2!} = \frac{5\times4\times3\times2\times1}{(3\times2\times1)\times(2\times1)} = 10 $$

In general, if there are n children and r of them are girls, the number of combinations is

$$ {}_nC_r = \frac{{}_nP_r}{r!} = \frac{n!}{r!(n-r)!} $$

7.3 Bayes theorem

Recall that

$$ P(A \text{ and } B) = P(A) \times P(B|A) $$

or, alternatively,

$$ P(B|A) = \frac{P(A \text{ and } B)}{P(A)} $$

This can be expanded to be written as

$$ P(B|A) = \frac{P(A|B) \times P(B)}{P(A)} = \frac{P(A|B) \times P(B)}{P(A|B) \times P(B) + P(A|\overline{B}) \times P(\overline{B})} $$

which is known as *Bayes theorem*.[6] Armed with this theorem, we can now answer the following question: given that the second birth was a girl, what is the probability that the first birth was a boy, that is, what is $P(B_1|G_2)$? Noting that the event $\overline{B}_1$ is G_1, Bayes theorem gives us

$$ P(B_1|G_2) = \frac{P(G_2|B_1) \times P(B_1)}{P(G_2|B_1) \times P(B_1) + P(G_2|G_1) \times P(G_1)} $$

$$ = \frac{0.4 \times 0.5}{(0.4 \times 0.5) + (0.6 \times 0.5)} $$

$$ = 0.4 $$

Thus, knowing that the second child was a girl allows us to *update* our probability that the first child was a boy from the *unconditional* value of 0.5 to the new value of 0.4.[7]

7.4 Definitions of probability

In our development of probability, we have ignored one important question: where do the actual probabilities come from? This, in fact, is a question that has vexed logicians and philosophers for several centuries and, consequently, there are (at least) three definitions of probability in common use today.

The first is the *classical* or *a priori* definition, and is the one that has been implicitly used in the illustrative examples. Basically, it assumes that each outcome in the sample space is *equally likely*, so that the probability that an event occurs is calculated by dividing the number of outcomes that indicate the event by the total number of possible outcomes in the sample space. Thus, if the experiment is rolling a fair six-sided die, and the event is throwing an even number, then the number of outcomes indicating the event is three (the outcomes $[2,4,6]$), and the total number of outcomes in the sample space is six $[1,2,3,4,5,6]$, so that the required probability is $3/6 = 1/2$.

This will only work if, in our example, we can legitimately assume that the die is fair. What would be the probability of obtaining an even number if the die was biased, but in an unknown way? For this, we need to use the *relative frequency* definition. If we conduct an infinite number of trials of an experiment (that is, $n \to \infty$), and the number of times an event occurs in these n trials is k, then the probability of the event is defined as the limiting value of the ratio k/n.

Suppose that even numbers were twice as likely to occur on a throw of the die as odd numbers (the probability of any even number is $2/9$, the probability of any odd number is $1/9$, so that the probability of an even number being thrown is $2/3$). However, we do not know this, and decide to estimate this probability by throwing the die a large number of times, say 10,000, and recording as we go the relative frequency of even numbers being thrown. A plot of this 'cumulative' relative frequency is shown in Figure 7.1.

We observe that it eventually 'settles down' on $2/3$, but it takes quite a large number of throws (around 6,000) before we get a really good estimate of the probability.

What happens if the experiment cannot be conducted more than once, if at all, for example, if it is a 'thought experiment'? We can then use *subjective probability*, which assigns a *degree of belief* to an event actually

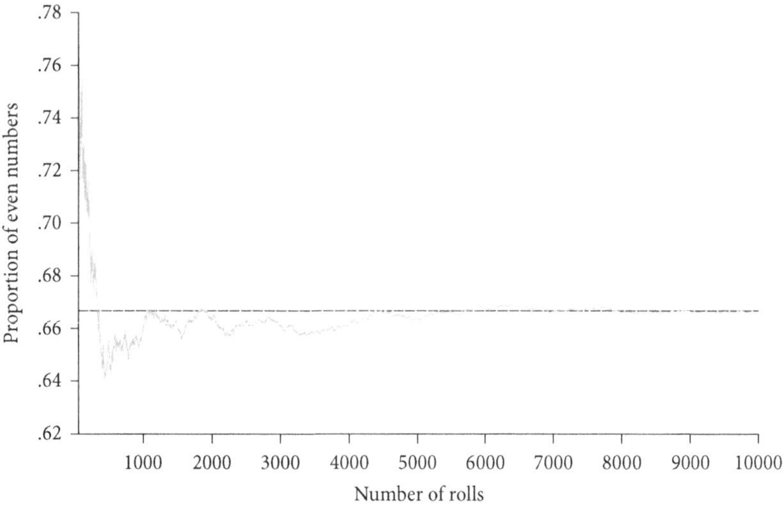

FIGURE 7.1 *Cumulative relative frequency plot for rolling a biased dice*

occurring. If we believe that it is certain to occur, then we assign a probability of one to the event; if we believe that it is impossible to occur then we assign a probability of zero; and if we think that it has a 'good chance' of occurring then we presumably assign a probability that is greater than 0.5 but less than 1, etc.

No matter what definition seems appropriate to the experiment and outcome at hand, fortunately all definitions follow the same probability axioms, and hence rules, as those outlined above.

Notes

1 The importance of exploratory data analysis in general was established with the publication of John W. Tukey's extraordinary book, *Exploratory Data Analysis* (Addison-Wesley, 1977), which remains a classic and is well worth reading.

2 There are 193 members of the United Nations. The US State Department actually recognises 195, but this does not include, for historical political reasons, Taiwan. Table 2.1 also includes data for Bermuda and Puerto Rica, which some authorities claim are not technically countries at all, the former being a territory of the UK, the latter a territory of the US. On the other hand, there are 209 countries in the FIFA world football rankings, which some may feel is a more comprehensive listing!

3 There is another, much more sophisticated, problem with time series data: what we actually observe over time is, in fact, just one possible *realisation* from a *stochastic process* (that is, a process whose successive values occur randomly). The time path of inflation could, if other events had occurred, have thus evolved differently to that which we actually observed. This is far too subtle a concept to discuss further here, but see, for example, Terence C. Mills, *Time Series Techniques for Economists* (Cambridge University Press, 1990).

4 That the basics of probability are typically explained using examples from games of chance (and we are no exception here) reflects the fact that the early concepts of probability theory were developed through the analysis of such parlour games. For example, if a pair of dice are rolled 12 times in succession, what should one bet on the chance of seeing at least one double six? How many rolls of the dice are required before the odds of seeing a double six is 50–50? Questions like these began to be asked around 1650 and attracted the attention of mathematicians such as Pascal and Fermat, who actually resolved the latter problem in what became the first theorem in probability. For more on the early history of probability, see Anders Hald, *A History of Probability and Statistics and Their Applications before 1750* (Wiley, 2005).

5 See Mike Barrow, *Statistics for Economics, Accounting and Business Studies*, 6th edition (Prentice Hall, 2013), chapter 3, for a brief discussion of tree diagrams.

6 The eponymous Bayes in the theorem is the Reverend Thomas Bayes (1702(?)–1761), an English dissenting minister who lived in Tunbridge Wells from 1731. His friend Richard Price found the theorem in Bayes' papers after his death and arranged for its posthumous publication ('An essay towards solving a problem in the doctrine of chances', *Philosophical Transactions of the Royal Society* 53 (1763)). For biographical details of Bayes, see Andrew I. Dale, *Most Honourable Remembrance: The Life and Work of Thomas Bayes* (Springer, 2003) and David R. Bellhouse, 'The Reverend Thomas Bayes, FRS: a biography to celebrate the tercentenary of his birth', *Statistical Science* 19 (2007), 3–43. That Bayes theorem might have been discovered earlier is discussed in Stephen M. Stigler, 'Who discovered Bayes's theorem?' *American Statistician* 37 (1983), 290–296.

7 Bayes' theorem, or 'rule' as it is often referred to, has since become the foundation for a very influential school of statistical analysis, that of *Bayesian inference*, an approach that is rather too advanced to be covered in this text. Dale J. Poirier, *Intermediate Statistics and Econometrics* (MIT Press, 1995), and Gary Koop, *Bayesian Econometrics* (Wiley, 2003), both provide introductory discussions of this very important approach to statistical modelling. The story of Bayes' theory through the last 250 years is recounted and

popularised in Sharon Bertsch McGrayne, *The Theory That Would Not Die: How Bayes' Rule Cracked the Enigma Code, Hunted Down Russian Submarines, & Emerged Triumphant from Two Centuries of Controversy* (Yale University Press, 2011). It is also a key technique for Nate Silver, *The Signal and the Noise: The Art and Science of Prediction* (Allen Lane, 2012).

8
Probability Distributions

Abstract: *A new concept, the random variable, is introduced to exploit fully the power of probability. Discrete random variables, those that take only a finite number of values, are focused upon, along with their associated* probability *distributions, which are essentially a listing of the values the random variable can take accompanied by their probabilities of occurrence. The concepts of expected value and variance of a random variable are developed. Some particular probability distributions are introduced that relate to types of probability experiments that occur in a variety of situations, most notably the binomial and Poisson (the latter being interpreted as an approximation to the former).*

8.1 Random variables

To be able to exploit fully the power of probability, we must introduce a new concept, that of the *random variable*. We have already encountered several examples of this: the outcome of a toss of a coin, a roll of a die or a draw of a card, and the number of girls in a family of five children, are all random variables. A random variable is therefore one whose outcome is the result of chance and is thus unpredictable. But we must be clear what we mean here by unpredictability: it does not mean that we know absolutely nothing at all about the values that a random variable can take; rather, it means that although we might know the values that can be taken, those values cannot be predicted to occur as an outcome with complete certainty. We know that when a coin is tossed it will land either heads or tails, but before the toss is made we do not know what the outcome will be (unless, of course, it is a two-headed coin!) Similarly with rolling a die and picking a card: the possible outcomes are determined by the sample space of the experiment, but none of them are *individually* certain to occur in a single trial of the experiment (although *one* of them will) and thus they all have probabilities associated with them. For example, rolling a 6 will occur with probability $1/6$ if the die is fair, but it is impossible to roll a 0 or a 7, which will thus have zero probabilities associated with them.

These are all examples of *discrete* random variables, where the sample space is defined over a finite number of outcomes. Many random variables have sample spaces associated with them that have an infinite number of outcomes, and these are known as *continuous* random variables. An example would be the height of a student drawn from those taking a particular module. In principle, this height could be measured to any degree of accuracy and thus could take on an infinite number of values. Of course, it may be argued that in practice measuring instruments are limited in their precision, so that we can only measure height over a finite number of values. A response to this would be that, although finite, the number of values could nevertheless be extremely large, so that we may as well act *as if* the random variable was continuous, and this is something that is done regularly in statistical analysis.

8.2 Probability distributions

As we have seen, the values or outcomes taken by a discrete random variable will have probabilities associated with them. A listing of these values and accompanying probabilities is called a *probability distribution*. For example, consider the probability distribution of the random variable defined to be the outcome of rolling the biased die of §7.4, in which even numbers were twice as likely as odd numbers: if the random variable is denoted X, and defined as $X_i = i$, $i = 1, 2, \ldots, 6$, then we can list the probability distribution of X as

X_i	1	2	3	4	5	6
$P(X_i)$	$1/9$	$2/9$	$1/9$	$2/9$	$1/9$	$2/9$

This shows clearly that a probability distribution has the following properties, which follow directly from the probability axioms of §7.2:

$$0 \le P(X_i) \le 1 \qquad\qquad E(X) = \sum_{i=1}^{n} X_i P(X_i)$$

Expected value and variance

Just as we can compute the mean of a sample of data, we can also compute a (weighted) mean for a random variable X. This is called the *expected value*, denoted $E(X)$, and is defined as the weighted sum of the n values taken by X, with the weights given by the associated probabilities:

$$E(X) = \sum_{i=1}^{n} X_i P(X_i)$$

For the weighted die,

$$E(X) = 1 \times 1/9 + 2 \times 2/9 + 3 \times 1/9 + 4 \times 2/9 + 5 \times 1/9 + 6 \times 2/9 = 33/9 = 3\,2/3$$

This has the following interpretation: if we rolled this die a large number of times, recorded the results of the throws, and calculated the cumulative average, this will eventually settle down to $E(X) = 3\,2/3$. Thus the expected value is often referred to as the *mean* of X.

Similarly, we can define the *variance* of X as

$$V(X) = E(X - E(X))^2 = \sum_{i=1}^{n} (X_i - E(X))^2 P(X_i)$$

so, for our weighted die,

$$V(X) = \left(-2\tfrac{2}{3}\right)^2 \times \tfrac{1}{9} + \left(-1\tfrac{1}{3}\right)^2 \times \tfrac{2}{9} + \left(-\tfrac{2}{3}\right)^2 \times \tfrac{1}{9} + \left(\tfrac{1}{3}\right)^2 \times \tfrac{2}{9} + \left(1\tfrac{1}{3}\right)^2 \times \tfrac{1}{9} + \left(2\tfrac{2}{3}\right)^2 \times \tfrac{2}{9}$$
$$= 2\,{}^{8}\!/_{9}$$

Alternatively, we can note that

$$\begin{aligned} V(X) &= E(X - E(X))^2 = E\left[(X - E(X))(X - E(X))\right] \\ &= E\left[X^2 - 2XE(X) + (E(X))^2\right] = E(X^2) - 2(E(X))^2 + (E(X))^2 \\ &= E(X^2) - (E(X))^2 \\ &= \sum^{n} X_i^2 P(X_i) - (E(X))^2 \end{aligned}$$

where we use the result that $E[E(X)] = E(X)$, because $E(X)$ is, by definition, a single number, that is, a constant, and the expectation of a constant must be itself. Thus

$$\begin{aligned} V(X) &= 1^2 \times \tfrac{1}{9} + 2^2 \times \tfrac{2}{9} + 3^2 \times \tfrac{1}{9} + 4^2 \times \tfrac{2}{9} + 5^2 \times \tfrac{1}{9} + 6^2 \times \tfrac{2}{9} - \left(3\tfrac{2}{3}\right)^2 \\ &= \tfrac{147}{9} - \tfrac{121}{9} = \tfrac{26}{9} \\ &= 2\,{}^{8}\!/_{9} \end{aligned}$$

$E(X)$ and $V(X)$ may thus be interpreted as measures of the central tendency and dispersion of the probability distribution of X. As with sample variances, we may take the square root of $E(X)(X)$ to be the *standard deviation* of X: $SD(X) = \sqrt{V(X)}$. For the above distribution, $SD(X) = \sqrt{26}/3 = 1.70$.

We may be able to associate particular probability distributions with random variables. For example, a fair six-sided die would have the probability distribution

$$P(X = i) = \begin{cases} \dfrac{i}{6} & i = 1,2,\ldots,6 \\ 0 & \text{otherwise} \end{cases}$$

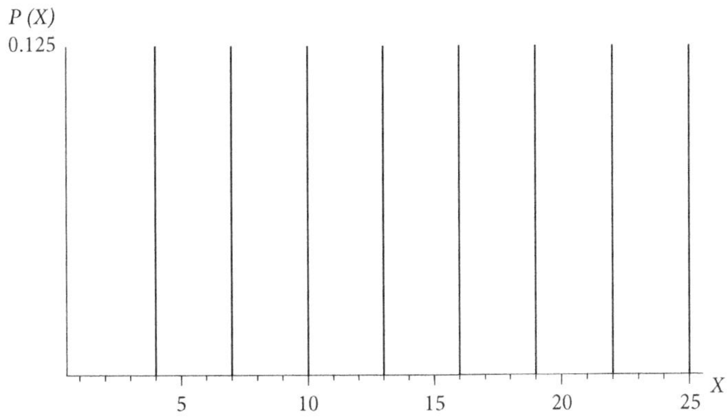

FIGURE 8.1 *A uniformly distributed random variable taking the values 4, 7, 10, 13, 16, 19, 22 and 25 with probability ⅛, with all other values having zero probability*

This is an example of the (discrete) *uniform distribution*, which can be written more generally as

$$P\left(X = k + hi\right) = \begin{cases} \frac{1}{n} & i = 1,2,\ldots,n \\ 0 & \text{otherwise} \end{cases}$$

and which can be shown to have expected value $E(X) = k + \frac{1}{2}h(n+1)$ and variance $V(X) = \frac{1}{12}h^2(n^2-1)$.[1] For $k = 1$, $h = 3$ and $n = 8$, so that X takes the values 4, 7, 10, 13, 16, 19, 22 and 25 with probability $\frac{1}{8}$ and all other values have zero probability, a graphical representation of the uniform distribution is shown in Figure 8.1: it has $E(X) = 1 + \frac{1}{2} \times 3 \times 9 = 14\frac{1}{2}$ and $V(X) = \frac{1}{12} \times 9 \times 63 = 47\frac{1}{4}$, so that $SD(X) = 6.87$.

For the fair die, $k = 0$, $h = 1$ and $n = 6$, and thus $E(X) = (n+1)/2 = 3\frac{1}{2}$ and $V(X) = (n^2 - 1)/12 = \frac{35}{12}$. When a variable X follows a uniform distribution with parameters k, h and n, we use the notation $X \sim U(k, h, n)$, which is to be read as 'X is distributed as uniform with parameters k, h and n'. The random variable defined to be the result of tossing a fair die is thus $X \sim U(0,1,6)$, and we can say that X is a uniformly distributed random variable.

Although simple, the uniform distribution lies behind most types of lottery and those situations where the only thing known about the outcome is that it can take one of a given set of values, so that all we can assume is that the probability of an outcome taking a particular value is the same for all values in the set.

8.3 The binomial distribution

A popular probability distribution defined for a discrete random variable is the *binomial*. This provides the general formula for computing the probability of r 'successes' from n independent trials of an experiment that has only two mutually exclusive outcomes, generically defined as 'success' and 'failure', and where the probability, p, of success (and hence the probability, $q = 1-p$, of failure), remains constant between trials.

As an example of a *binomial experiment*, recall the family of five children of which three are girls.[2] To calculate the probability of the event of three girls from five children, we have to be able to assume that the outcome of successive trials, that is, births, are independent and that the probability of a girl birth remains constant at p across trials (unlike the example in §7.2 in which girl births were not independent, so that the probability of such a birth was not constant). As we saw, there are ten possible orderings of three girls and two boys, and each of these will occur with probability $p^3(1-p)^2$, where we make use of the multiplication rule under independence. Thus, if X is the random variable defined to be the number of girls,

$$P(X=3)=10p^3(1-p)^2 = {}_5C_3 p^3(1-p)^2$$

Thus, if $p = 0.5$,

$$P(X=3)=10\times 0.5^3\times 0.5^2 = 0.3125$$

whereas if $p = 0.6$,

$$P(X=3)=10\times 0.6^3\times 0.4^2 = 0.3359$$

More generally, the probability of r successes from n trials is thus

$$P(X=r)={}_nC_r p^r(1-p)^{n-r}$$

It can be shown that $P(X=0)+P(X=1)+\ldots+P(X=n)=1$ and that $E(X) = np$ and $V(X) = np(1-p)$.[3] These formulae show that the binomial distribution has two parameters, the number of trials n and 'the' probability p, so that we can write $X \sim B(n,p)$ for a binomially distributed random variable.

Only allowing two outcomes is not as restrictive as it may seem. For example, we can compute the probability of, say, three 5 or 6s from six rolls of a fair die by defining the outcome of '5 or 6' as a success, with probability $p = \frac{1}{3}$, and 'not 5 or 6' as a failure, with probability $q = \frac{2}{3}$, that is, by defining $X \sim B(6,1/3)$ to be the number of 'successes'. The required probability is then

$$P(X=3) = {}_6C_3\left(\frac{1}{3}\right)^3\left(\frac{2}{3}\right)^3 = 20\times\frac{8}{729} = 0.2195$$

More complicated probabilities may be calculated by using the mutually exclusive version of the additive rule. For example, the probability of getting *more than* two girls in a family of five is given by

$$\begin{aligned} P(X>2) = P(X\geq 3) &= P(X=3)+P(X=4)+P(X=5) \\ &= 0.3125 + {}_5C_4\left(\frac{1}{2}\right)^5 + {}_5C_5\left(\frac{1}{2}\right)^5 \\ &= 0.3125 + 5\times\frac{1}{32} + 1\times\frac{1}{32} \\ &= 0.3125 + 0.15625 + 0.03125 = 0.5 \end{aligned}$$

Common uses of the binomial distribution include quality control, public opinion surveys, medical research, and insurance problems. It can be applied to complex processes such as sampling items in factory production lines or to estimate percentage failure rates of products and components.

8.4 The Poisson distribution

When the number of trials n is large and the probability of success p is small, another discrete distribution, the *Poisson*, provides an excellent approximation to the binomial. To see this, consider a typical example in industrial quality control. Here the number of trials (the number of items of a good that is produced) is large and, hopefully, the probability of a defective item is very small. Suppose that a manufacturer gives a two-year guarantee on the product that he makes, and from past experience knows that 0.5% of the items produced will be defective and will fail within the guarantee period. In a consignment of 500 items, the number of defectives will be distributed as $X \sim B(500, 0.005)$ and the probability that the consignment will contain no defectives is then 8.16%, calculated as

$$P(X=0) = {}_{500}C_0 \times 0.005^0 \times 0.995^{500} = 0.995^{500} = 0.0816$$

However, it turns out that, if n is large and p is small, such that $np \leq 7$, binomial probabilities are closely approximated by the values calculated using the following mathematical expression

$$P(X=r) = \frac{(np)^r e^{-np}}{r!} \qquad (e = 2.718\ldots) \qquad (8.1)$$

Thus, since $np = 500 \times 0.005 = 2.5$, we can approximate the probability of no defectives by

$$P(X=0) = \frac{2.5^0 e^{-2.5}}{0!} = e^{-2.5} = 0.0821 \qquad 0! \equiv 1$$

that is, 8.21%, which is probably accurate enough for most purposes.

This approximation becomes especially useful when more involved probabilities are required. The probability of more than three items being defective is given by

$$\begin{aligned} P(X>3) &= 1 - P(X=0) - P(X=1) - P(X=2) - P(X=3) \\ &= 1 - {}_{500}C_0 (0.995)^{500} - {}_{500}C_1 (0.005)(0.995)^{499} \\ &\quad - {}_{500}C_2 (0.005)^2 (0.995)^{498} - {}_{500}C_3 (0.005)^3 (0.995)^{497} \\ &= 1 - 0.0816 - 0.2050 - 0.2570 - 0.2144 \\ &= 0.2420 \end{aligned}$$

Using the above approximation, the calculation is much simpler and almost as accurate:

$$\begin{aligned} P(X>3) &= 1 - \frac{2.5^0 e^{-2.5}}{0!} - \frac{2.5^1 e^{-2.5}}{1!} - \frac{2.5^2 e^{-2.5}}{2!} - \frac{2.5^3 e^{-2.5}}{3!} \\ &= 1 - 0.0821 - 0.2052 - 0.2566 - 0.2138 \\ &= 0.2423 \end{aligned}$$

Equation (8.1) in fact provides the formula for the *Poisson distribution*, which is also known as the 'distribution of rare events'.[4] Since np is the expected value of the binomial distribution, with the Poisson we can denote this as μ, write (8.1) as

$$P(X=r) = \frac{\mu^r e^{-\mu}}{r!}$$

and say that $X \sim P(\mu)$, with $E(X) = \mu$. It must also be the case that the variance of the Poisson distribution will be $V(X) = \mu$, since with p small the variance of the binomial becomes $V(X) = np(1-p) \approx np = E(X)$. In this setup, there is no natural 'number' of trials: rather, we consider the number of trials to be the number of time or spatial intervals that the random variable is observed over, for which we know the mean number of occurrences per interval. Thus, suppose a football team scores an average of 1.5 goals per game and that we assume that the number of goals they score in any single game follows a Poisson distribution. Then the probability that the team scores no goals in a game is

$$P(X=0)=e^{-1.5}=0.2231$$

that is, about 8½ times in an English Premiership season of 38 games – whereas the probability that they score five goals in a game is

$$P(X=5)=\frac{1.5^5 e^{-1.5}}{5!}=0.0141$$

which is on average less than once a season!

The Poisson distribution has found application in the prediction of car accidents at traffic black spots, in modelling traffic flow and optimal 'gap distances', in predicting machine failures and in devising operating procedures in call centres.

8.5 Other related distributions

As we have seen, the Poisson provides a good approximation to the binomial when the number of trials is large and the probability of success is small. When the number of trials is small *and* successive trials can no longer be regarded as independent, then under certain forms of dependence we can use a *hypergeometric* distribution. If the trials are independent but there are more than two types of outcome, then we have a *multinomial* distribution. However, there is a very important distribution that results from another type of approximation to the binomial and it is to this that we turn to in the next chapter.

Notes

1 The expectation can be obtained straightforwardly:

$$\begin{aligned}E(X) &= (k+h)\tfrac{1}{n} + (k+2h)\tfrac{1}{n} + \ldots + (k+nh)\tfrac{1}{n}\\ &= \tfrac{1}{n}\big((k+h)+(k+2h)+\ldots+(k+nh)\big)\\ &= \tfrac{1}{n}\big(nk + h(1+2+\ldots+n)\big)\\ &= \tfrac{1}{n}\left(nk + h\tfrac{n(n+1)}{2}\right)\\ &= k + \tfrac{1}{2}h(n+1)\end{aligned}$$

where the standard summation formula $1 + 2 + \ldots + n = n(n+1)/2$ is used. Obtaining the variance is a little more complicated. We begin by obtaining $E(X^2)$:

$$\begin{aligned}E\left(X^2\right) &= (k+h)^2\tfrac{1}{n} + (k+2h)^2\tfrac{1}{n} + \ldots + (k+nh)^2\tfrac{1}{n}\\ &= \tfrac{1}{n}\left((k+h)^2 + (k+2h)^2 + \ldots + (k+nh)^2\right)\\ &= \tfrac{1}{n}\left(nk^2 + (1+2+\ldots+n)2hk + \left(1+2^2+\ldots+n^2\right)h^2\right)\\ &= k^2 + (n+1)hk + \tfrac{1}{6}(n+1)(2n+1)h^2\end{aligned}$$

where the further standard result is used. The variance is thus

$$V(X) = k^2 + (n+1)hk + \tfrac{1}{6}(n+1)(2n+1)h^2 - \left(k + \tfrac{1}{2}h(n+1)\right)^2$$

and some algebraic manipulation will produce $V(X) = \tfrac{1}{12}h^2\left(n^2 - 1\right)$.

2 This is often known as a *Bernoulli experiment* and the distribution a *Bernoulli distribution* after Jacob Bernoulli (1654–1705), who first introduced the concept, although strictly a Bernoulli experiment refers to just a single trial.

3 Derivations of these formulae are complicated and are omitted.

4 The distribution is so-named after Siméon Denis Poisson (1781–1840), who first published it in 1837. It was used by Ladislaus Bortkiewich (1868–1931), a Polish economist and statistician, who published a book about the distribution, titled *The Law of Small Numbers*, in 1898, where he noted that events with low frequency in a large population follow a Poisson distribution even when the probabilities of the events vary. In the book he gave an example of the number of soldiers killed by being kicked by a horse each year, showing that these numbers followed a Poisson distribution.

9

Continuous Random Variables and Probability Density Functions

Abstract: *An alternative way of approximating a binomial distribution is considered that leads to a continuous random variable (one that takes on an infinite number of values), known as the normal or Gaussian. Continuous random variables have probability density functions, rather than probability distributions, associated with them, and this leads to probabilities having to be calculated as an area under the function, which requires integral calculus. The standard normal distribution is introduced as a convenient way of calculating normal probabilities and examples of how to do such calculations are provided. Distributions related to the normal – the chi-square, Student's* t *and the* F *– along with the concepts of independence and covariance between random variables are introduced. Methods of simulating random variables and distributions are discussed.*

9.1 Another approximation to the binomial

Figures 9.1 to 9.3 provide graphical representations of a binomially distributed random variable X with $p = 0.4$, and with the number of trials increasing from $n = 6$, to $n = 20$, and finally to $n = 100$.

It is easy to calculate that, for $n = 6$, $E(X) = 2.4$ and $V(X) = 1.44$ for $n = 20$, $E(X) = 8$ and $V(X) = 4.8$, and for $n = 100$, $E(X) = 40$ and $V(X) = 24$, so that both the mean and the variance increase proportionately with n. Of more interest to us here is the skewness in the distributions. The coefficient of skewness for a binomial distribution is given by the formula $(1-2p)/\sqrt{np(1-p)}$ or $(1-2p)/SD(X)$, so that for $n = 6$, $skew = 1.83$, and the distribution is quite heavily skewed to the right, but for $n = 20$, $skew = 0.09$, so that the distribution is much more symmetric. Indeed, for $n = 100$, $skew = 0.04$ and, as well as being almost symmetric, the distribution is beginning to take on a distinct 'bell-like' shape.

This is an important feature of the binomial distribution, but note that as n increases, the number of values that X can take, r, also increases, and thus X becomes more and more like a *continuous* random variable. Figure 9.4 presents the plot of a function of a continuous random variable with

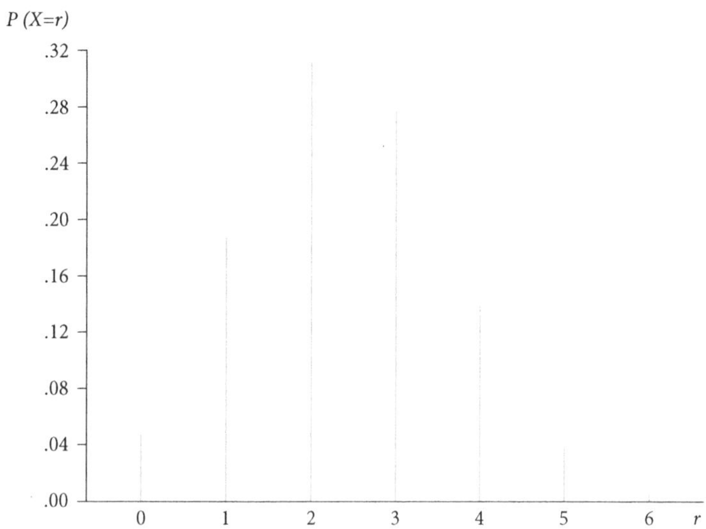

FIGURE 9.1 *Distribution of $X \sim B(6,0.4)$*

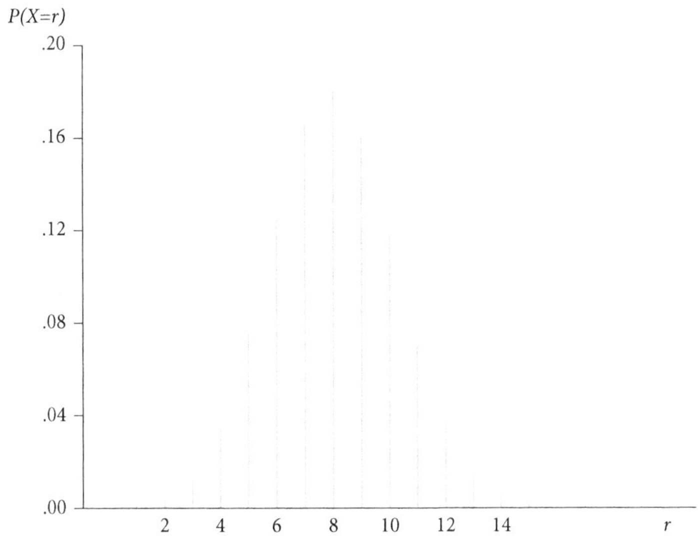

FIGURE 9.2 *Distribution of $X \sim B(20,0.4)$*

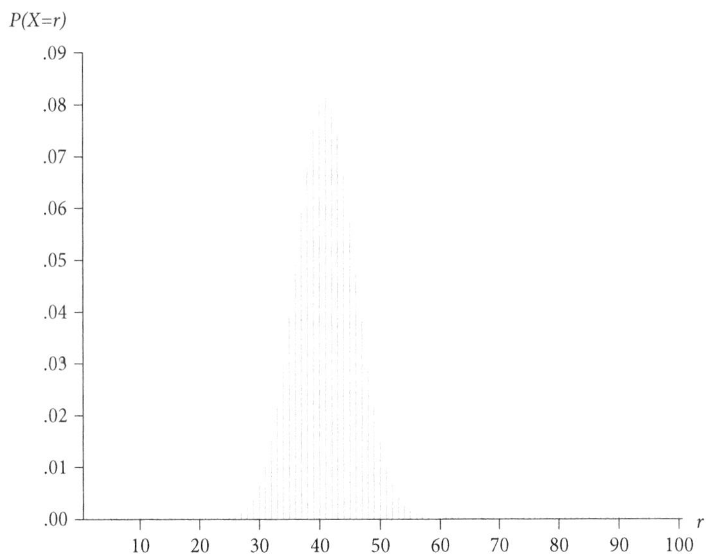

FIGURE 9.3 *Distribution of $X \sim B(100,0.4)$*

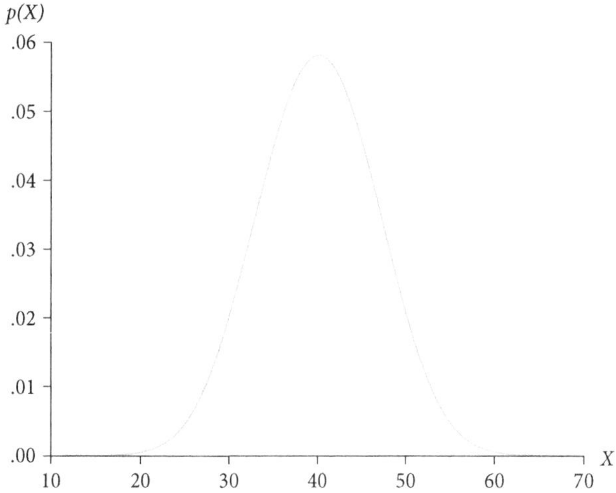

FIGURE 9.4 *Probability density function of* $X \sim N(40,24)$

the same mean, 40, and variance, 24, as the $B(100,0.4)$ discrete random variable shown in Figure 9.3.

The general form of the function, for mean μ and variance σ^2, is

$$p(X) = \frac{1}{\sqrt{2\pi\sigma^2}} \exp\left[-\frac{1}{2\sigma^2}(X-\mu)^2 \right] \quad -\infty < X < \infty \qquad (9.1)$$

and it defines a *normally distributed* random variable, denoted $X \sim N(\mu, \sigma^2)$. Thus, as the number of trials of a binomial experiment increases, the binomial distribution can be better and better approximated by a normal distribution having the same mean and variance (a good rule of thumb for the adequacy of the normal approximation is that $V(X) = np(1-p)$ should exceed 3). The normal distribution is bell shaped and symmetric about the mean μ, having its 'points of inflection' at $\mu - \sigma$ and $\mu + \sigma$, so that the standard deviation σ determines the 'width' of the distribution. The normal is perhaps the most famous distribution in statistics, with many physical phenomena appearing to be normally distributed. It was first proposed by the mathematician and astronomer, Karl Frederick Gauss, in 1809 as a model for the errors of measurement that occur in calculating the paths of stellar bodies, hence its alternative names, the *Gaussian distribution* and the 'normal curve of error'.[1]

9.2 Continuous random variables and calculating probabilities

As we can see from Figures 9.1 to 9.3, calculating the probability that a discrete random variable X equals a particular value r, that is, $P(X = r)$, is in effect carried out by simply reading the value on the vertical axis of the 'histogram' representation of the probability distribution corresponding to $X = r$. But with a continuous random variable, things are not so simple. Mathematically, since X now takes on an infinite number of values, the probability that it actually equals any particular value is infinitesimally small: in other words, $P(X = r) = 0$ for continuous X. Geometrically, the vertical 'line segment' at $X = r$ will have zero width and will cease to exist!

Obviously, we must be able to compute probabilities for continuous random variables, so what do we do? Although the probability that X *exactly* equals r is zero, the probability that X lies in an *interval around* r is non-zero, so that we can compute $P(r-e<X<r+e)$, which reads as 'the probability that X lies in the interval $r-e$ to $r+e$'. Geometrically, this will be given by the shaded area in Figure 9.5. Mathematically, it will be given by the integral of the function $p(X)$ evaluated between $r-e$ and $r+e$. The function $p(X)$ no longer has the interpretation of a probability distribution, but is now called the *probability density function* (pdf) of X. The form of the pdf defines the type of random variable that X is: if $p(X)$ takes the form (9.1) then X is a normally distributed random variable.

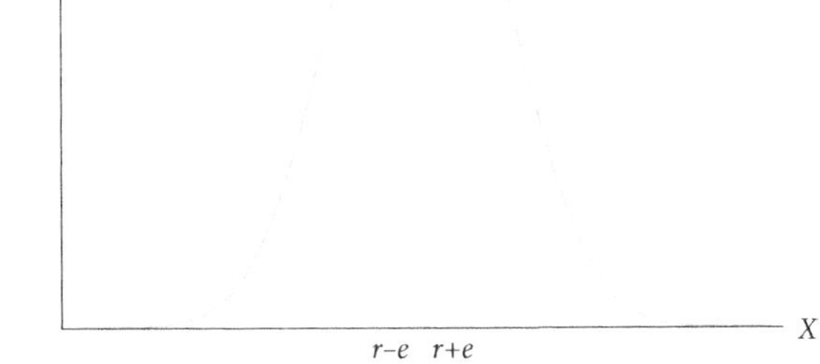

FIGURE 9.5 *Calculating probabilities from the normal probability density function*

9.3 The standard normal distribution

This 'area under a curve' interpretation of probability, for which we need methods of integral calculus, makes the computation of probabilities a much more complex problem to solve. It would appear that if we wanted to calculate the probability that a normally distributed random variable X falls in the interval a to b, then we would need to compute

$$P(a<X<b)=\frac{1}{\sqrt{2\pi\sigma^2}}\int_a^b \exp\left(-(X-\mu)^2/2\sigma^2\right)dX$$

Not only would we have to do this for every probability that we might wish to calculate, but a closed-form expression for the integral does not even exist! This looks to be an insurmountable problem, but help is at hand.

One of the properties of the normal distribution is that it is 'invariant to scale transformations', which means that if we multiply a normally distributed random variable by a constant and then add another constant to the result, we will still have a normal distribution, that is, if $X\sim N(\mu,\sigma^2)$, then $W=a+bX$ will also be normal. However, the mean and variance of W will not be the same as those of X:

$$E(W)=E(a+bX)=E(a)+E(bX)=a+bE(X)=a+b\mu$$

$$\begin{aligned}V(W)&=E(W-E(W))^2=E(a+bX-a+b\mu)^2=E(b(X-\mu))^2\\&=E\left(b^2(X-\mu)^2\right)=b^2E(X-\mu)^2=b^2\sigma^2\end{aligned}$$

Thus, if we set $a=-\mu/\sigma$ and $b=1/\sigma$, we can then define the new variable

$$Z=-\frac{\mu}{\sigma}+\frac{1}{\sigma}X=\frac{X-\mu}{\sigma}$$

which has

$$E(Z)=-\frac{\mu}{\sigma}+\frac{\mu}{\sigma}=0$$

and

$$V(Z)=\frac{\sigma^2}{\sigma^2}=1$$

Thus $Z \sim N(0,1)$ and is known as the *standard normal*, with pdf

$$p(Z) = \frac{1}{\sqrt{2\pi}} \exp\left(-Z^2/2\right)$$

This contains no unknown parameters and its integral can be evaluated using numerical procedures, so that the probability that Z falls into any particular interval can be computed straightforwardly (at least by using a computer). This enables probabilities for any normally distributed X to be calculated: since

$$P(a < X < b) = P\left(\frac{a-\mu}{\sigma} < Z < \frac{b-\mu}{\sigma}\right)$$

we can transform from X to Z and use the 'Z-probabilities' accordingly. Some examples will illustrate this procedure. Typical tables of what are known as the 'critical values' of the standard normal distribution give the values z_α such that $P(Z > z_\alpha) = \alpha$: they thus give 'right-hand tail areas', as shown in Figure 9.6.

Thus, for example, if $X \sim N(40,24)$ and we wish to find $P(42 < X < 50)$, then this probability is calculated as

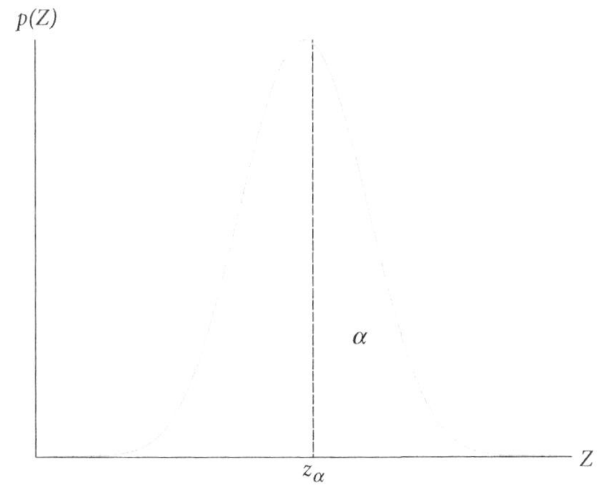

FIGURE 9.6 *Right-hand tail area of a normal distribution:* $P(Z > z_\alpha) = \alpha$

$$\begin{aligned} P(42 < X < 50) &= P\left(\frac{42-40}{\sqrt{24}} < Z < \frac{50-40}{\sqrt{24}}\right) = P(0.4082 < Z < 2.0412) \\ &= P(Z > 0.4082) - P(Z > 2.0412) \\ &= 0.3416 - 0.0206 = 0.3210 \end{aligned}$$

where $P(Z > 0.4082) = 0.3416$ and $P(Z > 2.0412) = 0.0206$ may be obtained from tables or appropriate software commands. Some useful 'Z-values' are those that leave 1%, 2.5%, 5% and 10% of the distribution in the right-hand tail (noting that $z_{0.50} = 0$ through symmetry):

$$z_{0.01} = 2.326 \qquad z_{0.025} = 1.960 \qquad z_{0.05} = 1.645 \qquad z_{0.10} = 1.282$$

Because of the symmetry of the normal distribution

$$P(Z < -z_\alpha) = P(Z > z_\alpha) = \alpha$$

and thus

$$P(Z > -z_\alpha) = 1 - P(Z < -z_\alpha) = 1 - P(Z > z_\alpha)$$

so that, for example,

$$\begin{aligned} P(30 < X < 42) &= P(-2.0412 < Z < 0.4082) \\ &= P(Z > -2.0412) - P(Z > 0.4082) \\ &= 1 - P(Z < -2.0412) - P(Z > 0.4082) \\ &= 1 - P(Z > 2.0412) - P(Z > 0.4082) \\ &= 1 - 0.0206 - 0.3416 \\ &= 0.6378 \end{aligned}$$

9.4 Distributions related to the normal

Several distributions that are in common use in economics and, in particular, in econometrics, are related to, and can be derived from, the normal. We must, however, first introduce the concept of *independent* random variables and the related idea of the *covariance* between two random variables. Recall from §7.2 that two events A and B are independent if $P(A \text{ and } B) = P(A)P(B)$. Now consider two random variables X and Y with probability density functions $p(X)$ and $p(Y)$, and let us introduce the *joint probability density function* $p(X,Y)$, which allows us to compute probabilities such as $P(a<X<b, c<Y<d)$. If $p(X,Y) = p(X)p(Y)$, so that $P(a<X<b, c<Y<d) = P(a<X<b)P(c<Y<d)$, then X and Y are said to be independent: if this is not the case then they are dependent.

Now recall the concept of the sample covariance from §5.2. The covariance between two random variables X and Y is analogously defined as

$$Cov(X,Y) = E(X - E(X))(Y - E(Y)) = E(XY) - E(X)E(Y)$$

If X and Y are both normally distributed then independence implies that they also have zero covariance, and vice versa, but this implication *does not* hold for variables that are not normally distributed, as these can be dependent and yet have zero covariance: recall the example of the circular relationship between two variables in §5.5.[2]

Let us now consider ν independent normal random variables, $X_1, X_2, \ldots, X_\nu$, with means $E(X_i) = \mu_i$ and variances $V(X_i) = \sigma_i^2$, $i = 1, 2, \ldots, \nu$, that is, $X_i \sim N(\mu_i, \sigma_i^2)$. If we define the random variable U as a weighted average (or linear combination) of the X_is

$$U = \sum_{i=1}^{\nu} a_i X_i = a_1 X_1 + a_2 X_2 + \ldots + a_\nu X_\nu$$

then $U \sim N(\mu_U, \sigma_U^2)$, where

$$\mu_U = \sum_{i=1}^{\nu} a_i \mu_i \qquad \sigma_U^2 = \sum_{i=1}^{\nu} a_i^2 \sigma_i^2$$

that is, a linear combination of normal random variables is also normally distributed, with its mean given by the same linear combination of the individual means and its variance given by a linear combination of the individual variances but with *squared* weights.

A special case of this result is that if $X_1 \sim N(0,1)$ and $X_2 \sim N(0,1)$ then clearly $X_1 \pm X_2 \sim N(0,2)$. A further special case is that if X_1 and X_2 are both $N(\mu,\sigma^2)$ then clearly $a_1 X_1 + a_2 X_2 \sim N\left((a_1 + a_2)\mu, (a_1^2 + a_2^2)\sigma^2\right)$ and, on using the scale transformation result of §9.3, it follows that

$$a_1 X_1 + a_2 X_2 = \sqrt{a_1^2 + a_2^2}\, X_3 + \left(a_1 + a_2 - \sqrt{a_1^2 + a_2^2}\right)\mu$$

where X_3 is also $N(\mu,\sigma^2)$.

The converse of this result is also true: if $X_1, X_2, \ldots, X_\nu$ are independent and their sum is normal then they must individually be normal. If there are two linear combinations $U_1 = \sum a_i X_i$ and $U_2 = \sum b_i X_i$ then these will be independent if and only if $\sum a_i b_i \sigma_i^2 = 0$.

The chi-square distribution

Now consider the standard normal transform $Z_i = (X_i - \mu_i)/\sigma_i^2$ and construct the 'sum of squares'

$$\chi^2 = \sum_{i=1}^{\nu} Z_i^2 = Z_1^2 + Z_2^2 + \ldots + Z_\nu^2$$

The random variable χ^2 then has a χ^2 (chi-square) distribution with ν *degrees of freedom*: $\chi^2 \sim \chi^2(\nu)$.As it is defined as a sum of squares, $\chi^2 > 0$, and so the variable can take only positive values. The χ^2 distribution is, to be precise, a 'family' of distributions in which the shape of the pdf (which takes a very complex form that need not concern us here) depends on ν, as do the mean and variance: $E(X) = \nu$, $V(X) = 2\nu$.

Figure 9.7 shows the pdfs for $\chi^2 = 1$ and 10.

Since the coefficient of skewness for a χ^2 variable is $2\sqrt{2/\nu}$, this takes the values 2.83 and 0.89 respectively, so that the χ^2 distribution becomes more symmetric as the degrees of freedom increase: in fact, for large ν it is approximately $N(\nu, 2\nu)$.

The Student's t distribution

If Z is a standard normal random variable that is independent of χ^2, then the random variable defined as

$$T = \frac{Z}{\sqrt{\chi^2/\nu}}$$

is said to have a (Student's) t distribution with ν degrees of freedom: $T \sim t(\nu)$.[3] The t distribution has $E(T) = 0$ and $V(T) = \nu/(\nu-2)$, and is thus only defined for $\nu > 2$, otherwise the variance would not be positive. It is also symmetric and, since $V(T) \rightarrow 1$ as $\nu \rightarrow \infty$, the family of t distributions converges to a standard normal, although for finite ν it will always have a larger variance than Z, so that it will always have 'fatter tails' than the standard normal.

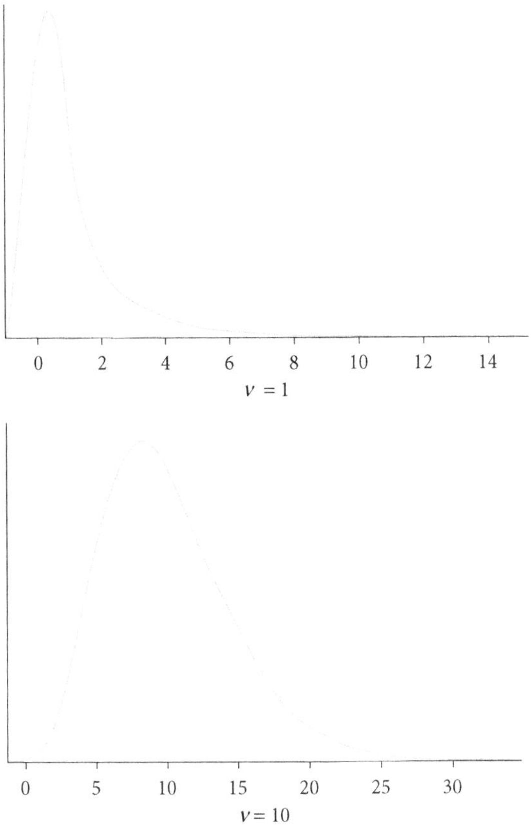

FIGURE 9.7 $\chi^2(\nu)$ *probability density functions*

Figure 9.8 shows the pdfs for $t(3)$ and $t(20)$ random variables, with the standard normal shown for comparison.

The fatter tails of the t distribution are best seen by comparing the values that leave, say, 5% in the right hand tail of the distribution: $t_{0.05}(3) = 2.353$ and $t_{0.05}(20) = 1.725$ compared to $z_{0.05} = 1.645$. For any α, the $t_\alpha(\nu)$ values converge to z_α as $\nu \rightarrow \infty$.

The F distribution

Suppose we have two independent chi-square random variables, $\chi_1^2 \sim \chi^2(\nu_1)$ and $\chi_2^2 \sim \chi^2(\nu_2)$. The random variable defined as the ratio of the two (scaled by their degrees of freedom)

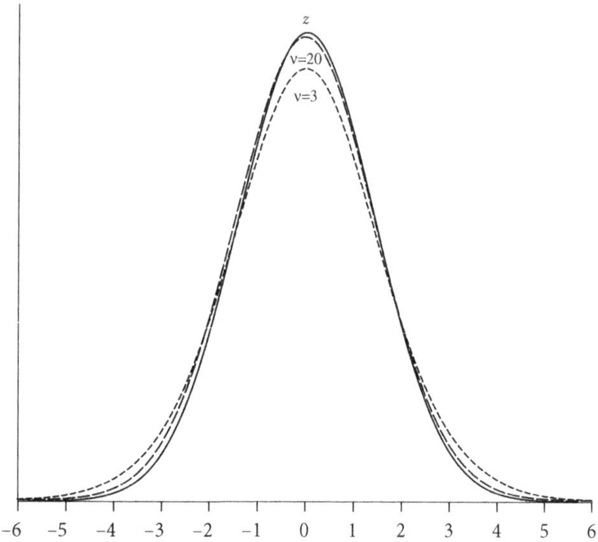

FIGURE 9.8 *$t(\nu)$ probability density functions compared to the standard normal z*

$$F = \frac{\chi_1^2 / \nu_1}{\chi_2^2 / \nu_2} \tag{9.2}$$

then has an F distribution with ν_1 numerator degrees of freedom and ν_2 denominator degrees of freedom, that is, $F \sim F(\nu_1, \nu_2)$.

Typically in economic applications ν_1 is a lot less than ν_2, so that Figure 9.9 shows the pdf of an $F(3, 50)$ random variable.

It has the right skewness typical of F distributions. Two results concerning F distributions can prove useful in obtaining probabilities. From the definition (9.2) it follows that

$$F_{1-\alpha}(\nu_2, \nu_1) = \frac{1}{F_\alpha(\nu_1, \nu_2)}$$

which enables left-hand tail critical values to be obtained from right-hand tail critical values. There is also a link between the F and t distributions through

$$F(1, \nu_2) = t^2(\nu_2)$$

FIGURE 9.9 *F*(3,50) *probability density function*

that is, the square of a *t* distribution with ν_2 degrees of freedom is an *F* distribution with 1 and ν_2 degrees of freedom.

9.5 Simulating distributions

It is often useful to simulate distributions using a *random number generator*. Figure 9.10 shows 500 simulated *U*(0,1) and *N*(0,1) variables: in effect, 500 random drawings have been made from the two distributions, and these have then been plotted as a pair of time series.[4]

The 'side bars' to the plots show the histograms of the simulated distributions. Even though these 'sample distributions' (a concept to be introduced formally in §10.2) seem to contain rather a large number of values, the histograms only provide approximations to the uniform and normal distributions from which the samples are drawn from. For example, the mean and standard deviation of the *U*(0,1) distribution are 0.5 and 0.2887 respectively, while the sample counterparts here are 0.5047 and 0.2878. Similarly, the sample mean and standard deviation of the simulated standard normal are 0.0535 and 1.0033.

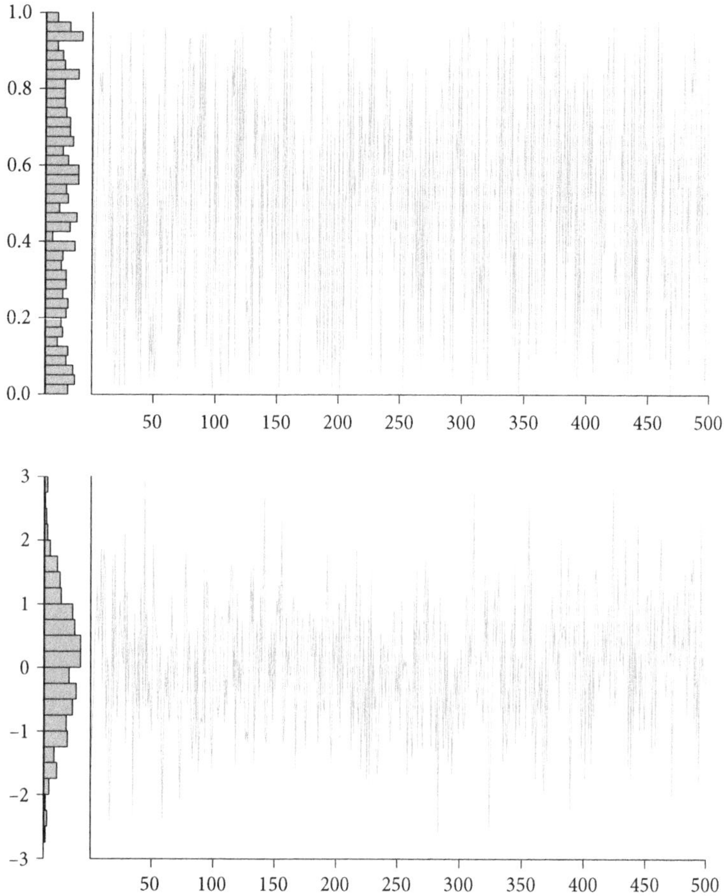

FIGURE 9.10 *Simulations of 500 drawings from U(0,1) (top) and N(0,1) (bottom) distributions*

Note that the maximum and minimum simulated values of the standard normal are 2.938 and −2.674: the probability that a standard normal lies outside the range $-3 < X < 3$ is 0.0027, which we should expect to occur only about once every 500 draws.[5]

Notes

1 A convenient source of information on the normal distribution, containing much more than is required here, is the Wikipedia entry http://en.wikipedia.

org/wiki/ Normal_distribution. An interesting account of how the distribution became 'normal' and how a particular form of it became 'standard' (see §9.3) is given by Stephen M. Stigler, 'Normative terminology' (with William M. Kruskal), chapter 22 of *Statistics on the Table* (Harvard, 1999).

2 Technically, independence requires that the joint probability distribution $p(X,Y)$ be equal to the product of the marginal probability distributions, $p(X)$ $p(Y)$, which will imply that so that $Cov(X,Y) = 0$. However, the converse will not hold except in the case of jointly normally distributed random variables, which are completely defined by $E(X)$, $E(Y)$ and $E(XY)$.

3 The rather unusual name, 'Student', given to this distribution is a consequence of it being proposed by William Sealy Gosset (1876–1937), one of the most famous statisticians of the 20th century: see Student, 'The probable error of the mean', *Biometrika* 6 (1908), 1–25. As an employee of Guinness, the famous Irish brewer of stout, Gosset was precluded from publishing under his own name and thus took the above pseudonym: see 'That dear Mr Gosset', chapter 3 of David Salsburg, *The Lady Tasting Tea: How Statistics Revolutionized Science in the Twentieth Century* (New York, Henry Holt, 2001).

4 The first simulation actually uses the *continuous* uniform distribution $U(a,b)$, rather than the discrete version introduced in §8.2, hence the slight alteration in notation. This has the pdf $p(x) = (b - a)^{-1}$ for $a \leq x \leq b$, $b > a$, with mean $\frac{1}{2}(a + b)$ and variance $\frac{1}{12}(b-a)^2$. The uniform distribution plays a key role in modern *random number generation* (RNG), as random number generators for many other distributions can be derived from the uniform generator: see Jurgen A. Doornik, 'The role of simulation in econometrics', in Terence C. Mills and Kerry Patterson (editors), *The Palgrave Handbook of Econometrics, Volume 1: Econometric Theory* (Palgrave Macmillan, 2006), pp. 787–811.

5 An extension of this idea lies behind the '6 sigma' business management strategy originally introduced by Motorola in the 1980s; only 3.4 per million draws from a normal distribution will be more than 6σ away from the mean.

10

Sampling and Sampling Distributions

Abstract: *The issue of sampling from an underlying population is considered more formally, with the distinction being drawn between deductive and inductive statistical reasoning. To allow the ideas of statistical inference to be analysed, the concept of a simple random sample is introduced, along with the related ideas of accuracy and precision. The sampling distribution of the mean from a normal population is developed and the result extended, through the central limit theorem, to non-normal populations. The sampling distribution of the variance is then considered.*

10.1 Sampling from a population

In §7.1 we argued that most, if not all, of the data sets used in earlier examples should be thought of as *samples* drawn, in some manner, from an underlying *population* which we cannot observe in its entirety. We now wish to return to this issue, and to consider the whole idea of sampling more formally.

To do this successfully, we must distinguish between deductive and inductive statistical reasoning. With *deductive* statistical reasoning, we start with a population about which we already know everything that we need or want to know, and we investigate the possible samples and their characteristics that could be obtained from our known population. This idea forms the basis of this section.

Inductive statistical reasoning, on the other hand, starts with a known sample and attempts to identify the population that actually generated the sample. Usually we will already know the population in a general way. What we won't know – or what we want to test – is the actual value of some specific population characteristic, for example, the mean. We have the value of the sample counterpart – the sample mean – so how do we use this information to *infer* the value of the population mean? Subsequent sections will explore this problem of *statistical inference* in some detail.

An obvious question is – why sample at all? As we have seen, often only samples of data are available and we have to make do with what we have got. This is particularly the case with historical, typically time series, data. Another important reason is that it is costly and time consuming to perform a complete enumeration of the population: hence censuses are typically taken only every decade and opinion polls are ever popular! In some cases, enumerating the population would be counterproductive. Consider a producer of light bulbs who wants to know their mean life. One way of doing this is to plug in every bulb produced and clock the time it takes for each to burn out. The producer would certainly end up with a very accurate estimate of the mean life of a light bulb, but unfortunately would also have no product left to sell! In such cases of *destructive sampling*, taking just a limited sample is the only viable way of obtaining such information.

A second question is – if a sample is taken, how precise is the resulting information? This question implies a host of other questions, the most fundamental being – *how* was the sample selected? There are two basic alternatives: a *judgement sample* and a *random sample.* At first sight, it may be thought that a judgement sample must be preferable to a random sample on the grounds that using one's judgement to select a sample must be better than choosing it completely at random. Nothing could be further from the truth, however! Judgement samples selected by an 'expert' are rife with implicit biases that we have no way of measuring, whereas random samples are based on procedures where the probability of getting each sample is known precisely.

A *simple random sample* is the result of selecting a sample from a population in such a way that *all* possible samples have the *same probability* of being selected. If there are K possible samples from a particular population, then each and every sample must have probability $1/K$ of being selected in order for the procedure to be regarded as simple random sampling. We shall only consider simple random sampling, but there are other random sampling techniques, such as *stratified sampling, cluster sampling* and *multi-stage sampling,* that are often used in more complex sampling problems.[1]

Whatever method of sampling is used, the relevant characteristic of the population, say the mean, will almost certainly be different from the same characteristic in the sample: the difference is called the *sampling error.* Random sampling allows us to measure the extent of sampling error. There will usually be another type of error, which is unrelated to sampling. This is called *non-sampling,* or *systematic, error* and results from such things as ill-defined questionnaires, in the form of vague or ambiguous questions, and incorrectly calibrated measuring devices.

These ideas about errors can be formalised to allow a distinction to be made between *accuracy* and *precision.* Let us consider using a sample mean $\bar{X}$ to provide an estimate of the unknown population mean μ. We can then write the 'decomposition'

$$\mu = \bar{X} + \varepsilon_s + \varepsilon_{ns}$$

where ε_s is the sampling error and ε_{ns} is the non-sampling error. If we took a census, ε_s would be zero, but there is no guarantee that ε_{ns} would be zero. When ε_{ns} is zero (or essentially zero) we say that the measurement is *accurate.* When ε_s is as small as we want it to be, we say that the measurement is *precise.* The sample can only provide us with information on precision, with statistical inference concentrating on the extent of the sampling error. Accuracy is a function of 'non-statistical' *experimental design*, which is not considered here, and hence we assume that all measurements in this chapter are accurate.

10.2 The sampling distribution of the mean

Let us assume that the population of the random variable X that we are interested in is normally distributed: $X \sim N(\mu,\sigma^2)$. The mean μ and variance σ^2 are the *parameters* of the distribution, whose actual values are unknown. We wish to *estimate* these parameters by drawing a simple random sample of size N. We denote these sample values as $X_1, X_2, \ldots, X_N$. From these values we can calculate the sample mean $\bar{X}$ and sample variance σ^2. These are known as *(sample) statistics*, and they can then be used to make inferences (estimates or decisions) about the population parameters.

A crucial point to understand is that any random sample of N values has occurred by chance – we could have obtained a completely different set of values. The sampled values $X_1, X_2, \ldots, X_N$ must therefore all be random variables, so that any sample statistic constructed from them must also be a random variable, whose value would change from sample to sample. Being random variables, sample statistics will have distributions associated with them, known as *sampling distributions*. To be able to make inferences about population parameters from sample statistics requires knowledge of the statistic's sampling distribution.

Let us concentrate on the sample mean, $\bar{X}$, which can be thought of as a linear combination of the sample values:

$$\bar{X} = \frac{1}{N}\sum_{i=1}^{N} X_i = \frac{1}{N}X_1 + \frac{1}{N}X_2 + \ldots + \frac{1}{N}X_N$$

In the notation of §9.4, $a_i = 1/N$ for $i = 1, \ldots, N$. One of the properties of simple random sampling is that, since each sampled value is equally likely to occur, successive values are independent of each other. Moreover, since each value is drawn from the same normal distribution, the X_i are normally distributed with common mean μ and common variance σ^2. Hence, using the results on linear combinations of independent random variables in §9.4,

$$E\,\bar{X} = \mu_{\bar{X}} = \frac{1}{N}\mu + \frac{1}{N}\mu + \ldots + \frac{1}{N}\mu = \frac{N}{N}\mu = \mu$$

$$V(\bar{X}) = \sigma_{\bar{X}}^2 = \frac{1}{N^2}\sigma^2 + \frac{1}{N^2}\sigma^2 + \ldots + \frac{1}{N^2}\sigma^2 = \frac{N}{N^2}\sigma^2 = \frac{\sigma^2}{N}$$

and, because a linear combination of normal random variables is also normal, the sampling distribution of the sample mean is thus

$$\bar{X} \sim N\left(\mu, \sigma^2/N\right)$$

Hence, it follows that

$$Z = \frac{\bar{X} - \mu}{\sigma/\sqrt{N}} = \sqrt{N}\left(\frac{\bar{X} - \mu}{\sigma}\right) \sim N(0,1)$$

Figure 10.1 shows a population that follows a $N(100,20)$ distribution, and the distribution of a sample of size 10 drawn from this population, which will be $N(100,2)$.

Intuitively, if the population mean is 100, then a random sample of size $N = 10$ is expected to yield a sample mean of around the same value, perhaps a little more, perhaps a little less. However, while it is quite likely that any single individual drawn from this population will be, say, at least 105

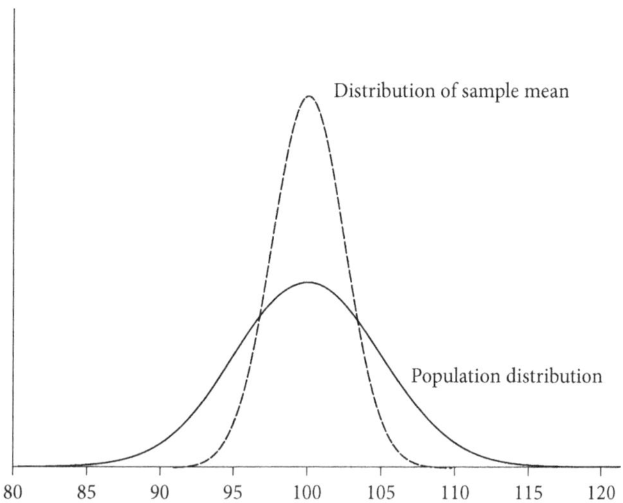

FIGURE 10.1 *Normal population and the distribution of the sample mean*

(the probability is $P(X>105)=P\left(Z>(105-100)/\sqrt{20}=1.118\right)=0.132$, it is much less likely that the mean of 10 individuals drawn from the population will be at least 105, since the probability of this occurring is $P\left(\bar{X}>105\right)= P\left(Z>(105-100)/\sqrt{2}=3.536\right)=0.029$, that is, around 3% rather than 13%. This is because, in order to get a sample mean in excess of 105, we would need to draw approximately half the sample with values greater than this, which will be correspondingly less likely. By extension, the larger the sample, the smaller the variance of the sample mean. If the sample size is 100, $\bar{X}\sim N\left(100,0.2\right)$ and $P\left(\bar{X}>105\right)=P\left(Z>(105-100)/\sqrt{0.2}\right)=11.18$, which is essentially zero!

Note that the variance of the sample mean is $\sigma_{\bar{X}}^2=\sigma^2/N$. Its square root, $\sigma_{\bar{X}}=\sigma/\sqrt{N}$, is known as the *standard error*, to distinguish it from σ the standard deviation of the population.

10.3 Sampling from a non-normal population

The above analysis is based on the assumption that the population is normal. What happens if the population follows some other distribution? Extraordinary as it may seem, nothing – as long as the sample size is reasonably large! This is a consequence of one of the most important and useful results in statistics, the *central limit theorem* (CLT). This states that the distribution of the sample mean, calculated from a random sample, approaches $\bar{X}\sim N\left(\mu,\sigma^2/N\right)$ as $N\rightarrow\infty$ *irrespective* of the distribution of the population.[2] More formally, if $X_1,X_2,\ldots,X_N$ are independent and identically distributed with common mean μ and common variance σ^2 then, as $N\rightarrow\infty$,

$$\sqrt{N}\left(\bar{X}-\mu\right)\overset{d}{\sim}N\left(0,\sigma^2\right)$$

where 'd' denotes 'convergence in distribution'.

As an example of the CLT, consider Figure 10.2, which is the probability distribution of a discrete and asymmetric Poisson random variable with parameter 1.5, that is, the population is $X\sim P(1.5)$.

In §8.4 we stated that the mean of a Poisson distribution is given by its parameter, μ. We can now state that its variance is also given by μ and its coefficient of skewness by $1/\sqrt{\mu}$: hence $V(X)$ = 1.5 and $skew(X)$ = 0.82.

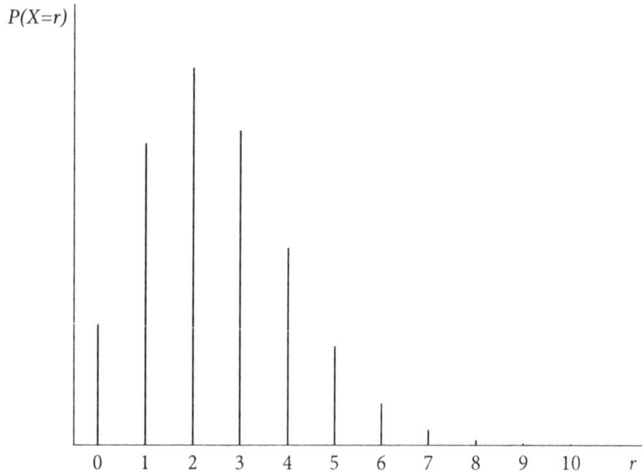

FIGURE 10.2 $X \sim P(1.5)$ *probability distribution*

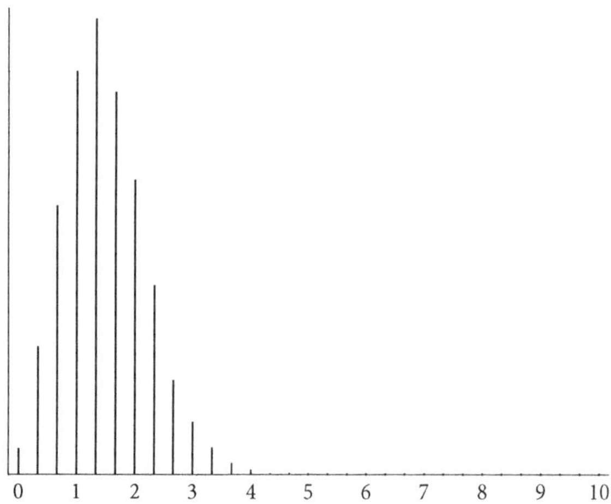

FIGURE 10.3 *Distribution of the mean of a sample of size* $N = 3$ *from* $X \sim P(1.5)$

Figure 10.3 shows the sampling distribution of $\bar{X}$ for $N = 3$. This is still a discrete distribution, taking the values $0, \frac{1}{3}, \frac{2}{3}, 1, 1\frac{1}{3}, 1\frac{2}{3}, 2, \ldots$, and has a mean of 1.5 and a standard error of $\sqrt{1.5/3} = 0.70$, but is seen to be less skewed than the population X distribution.

Figure 10.4 shows the sampling distributions of $\bar{X}$ for $N = 10$ and 25, and these can now be approximated by density functions.

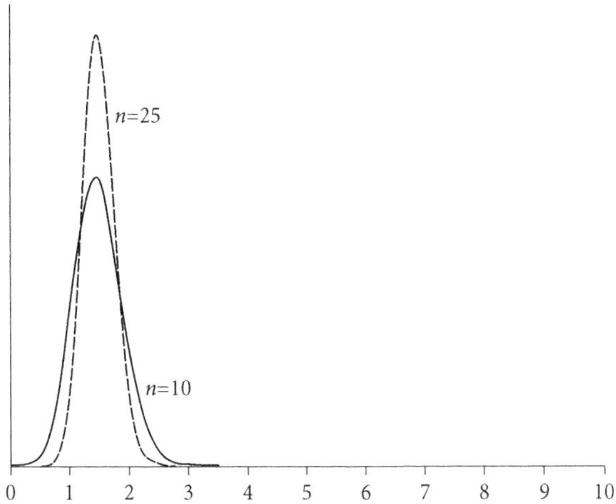

FIGURE 10.4 *Distribution of the mean from samples of size N = 10 and N = 25 from X ~ P(1.5)*

Even with $N = 10$, the density function is close to the normal, and it certainly is with $N = 25$, thus showing that even for asymmetric, discrete random variables, the CLT very quickly produces a normal distribution for the sample mean as the sample size increases.

10.4 The sampling distribution of the variance

Let us now consider the distribution of the sample variance calculated from a random sample of size N from an $X \sim N(\mu, \sigma^2)$ population. Since each of the values in the sample will have the same mean and variance, we can define the standard normal variables $Z_i = (X_i - \mu)/\sigma$ and recall from §9.4 that

$$\sum_{i=1}^{N} Z_i^2 = \frac{1}{\sigma^2}\sum_{i=1}^{N}(X_i - \mu)^2 \sim \chi^2(N)$$

If the unknown μ is replaced by the sample mean $\bar{X}$, then the corresponding random variable continues to be distributed as chi-square, but with one less degree of freedom:

$$\frac{1}{\sigma^2}\sum_{i=1}^{N}(X_i - \bar{X})^2 \sim \chi^2(N-1)$$

Now recall that the sample variance is

$$s^2 = \frac{1}{N-1}\sum_{i=1}^{N}\left(X_i - \bar{X}\right)^2 \qquad \text{i.e.,} \qquad \sum_{i=1}^{N}\left(X_i - \bar{X}\right)^2 = (N-1)s^2$$

Thus

$$(N-1)\frac{s^2}{\sigma^2} \sim \chi^2(N-1) \qquad \text{or} \qquad s^2 \sim \frac{\sigma^2}{(N-1)}\chi^2(N-1)$$

that is, the sample variance is distributed as a *scaled* chi-square variable with $N-1$ degrees of freedom, the scaling factor being $\sigma^2/(N-1)$.

Thus if the population is $N(100,20)$ and a sample of size $N = 10$ is taken,

$$s^2 \sim \frac{20}{9}\chi^2(9)$$

The sampling distribution of s^2 is shown in Figure 10.5. Suppose we want to calculate the probability that we observe a sample variance greater than 30. This is calculated as

$$P(s^2 > 30) = P\left(\chi^2(9) > 30 \times \tfrac{9}{20}\right) = P\left(\chi^2(9) > 13.5\right) = 0.141$$

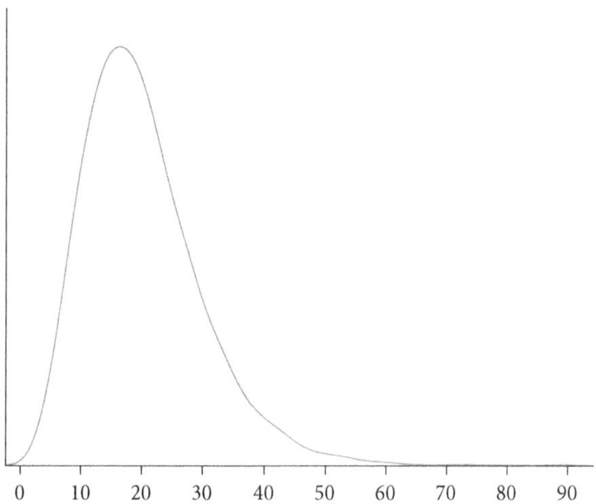

FIGURE 10.5 *Sampling distribution of* $s^2 \sim (20/9)\chi^2(9)$

On the other hand, the probability that the sample variance is less than 10 is

$$P(s^2 < 10) = P\left(\chi^2(9) < 10 \times \tfrac{9}{20}\right) = P\left(\chi^2(9) < 4.5\right) = 0.124$$

The difference in the two probabilities (which would be the same if the sampling distribution was symmetric) is a consequence of the lower bound of zero on s^2 inducing asymmetry into the distribution, which can clearly be seen in Figure 10.5.

Notes

1 The classic text on sampling remains William G. Cochran, *Sampling Techniques*, 3rd edition (Wiley, 1977). A recent general discussion is provided by Robert M. Groves, Floyd J. Fowler, Mick P. Couper, James M. Lepkowski, Eleanor Singer and Roger Tourangean, *Survey Methodology*, 2nd edition (Wiley, 2010).

2 There are various versions of the CLT; the one we focus on here is the *classical* or *Lindeberg–Lévy* version. Forms of the theorem go back at least to de Moivre in 1733, although the term 'central limit theorem' was first used only in 1920.

11
Estimation and Inference

Abstract: *The difference between an estimate and an estimator is emphasised and some properties of the latter, such as unbiasedness, consistency and efficiency, are introduced. The concepts of confidence intervals for the mean and variance are developed and their interpretation discussed by way of an example using income inequality data. Hypothesis testing is then introduced, and procedures for testing hypotheses about the mean and variance are proposed. Further considerations concerning hypothesis testing, such as Type I and II errors, power and prob-values, are discussed. These concepts are used to develop methods for performing inference on correlation coefficients, with a test for zero correlation and a confidence interval for the correlation coefficient being constructed.*

11.1 Estimators and their Properties

In §10.2, we focused attention on the sample mean $\bar{X}$ as a statistic with which to make inferences about the population mean μ. The value of $\bar{X}$ calculated from a particular random sample of size N is an *estimate* of μ, while the formula used for the calculation, $\bar{X} = \sum X / N$, is referred to as an *estimator*. The sample mean is not the only estimator of the population mean, for the sample median and mode are clearly alternatives. The sample mean does, though, have some desirable statistical properties that make it a popular choice. While we shall not dwell on these in detail here, recall that in §10.2 we showed that $E(\bar{X}) = \mu$: that the expected value of the distribution of the sample mean was the population mean. This is the property of *unbiasedness*. We also showed that the variance of the sampling mean was the ratio σ^2/N. Thus, as $N \to \infty$, this variance tends to zero, and this is the property of *consistency*, that is, as samples get larger and larger, the sampling distribution of the sample mean becomes more tightly packed around the population mean. The sample median can also be shown to be unbiased and consistent, so why should we focus on the sample mean as an estimator of the population mean, to the exclusion of the median? The reason is that the variance of the sample median can be shown to be $(\pi/2) \times (\sigma^2/N)$, so that it is approximately 1.57 times bigger than the variance of the sample mean.[1] The distribution of the sample median is thus more spread out than the corresponding distribution of the sample mean, so that we have a greater chance of getting a sample median estimate much further away from the population mean than we do if we use the sample mean. The sample mean is thus said to be *relatively efficient* compared to the sample median, and indeed it can be shown to be relatively efficient when compared to any other unbiased estimator of the population mean from a normal distribution.[2]

Not all estimators are unbiased. Suppose that we have an estimator $\hat{\theta}$ of a population parameter θ. If $E(\hat{\theta}) \neq \theta$ then $\hat{\theta}$ is *biased* and the bias is given by $\hat{\theta} - \theta$. Consider the estimator of the population variance σ^2 given by

$$s_*^2 = \frac{1}{N} \sum_{i=1}^{N} \left(X_i - \bar{X} \right)$$

that is, we use the divisor N rather than $N-1$. Following the same derivation as in §10.4,

$$N\frac{s_*^2}{\sigma^2}\sim\chi^2(N-1) \qquad \text{or} \qquad s_*^2\sim\frac{\sigma^2}{N}\chi^2(N-1)$$

Thus

$$E(s_*^2)=\frac{\sigma^2}{N}E\left(\chi^2(N-1)\right)=\frac{N-1}{N}\sigma^2<\sigma^2$$

and the bias is $-\sigma^2/N$.

It is straightforward to show that s^2, on the other hand, *is* unbiased. Note that as $N\rightarrow\infty$, the bias declines to zero, so that s_*^2 is *asymptotically unbiased*. It can also be shown to be a consistent estimator, as is s^2, so that for large N it does not matter whether we use s_*^2 or s^2, although for small N the latter should always be used.

11.2 A confidence interval for the population mean

Suppose that we are in the simple, but unrealistic, situation of having a population that we know to be normal and that we also know its variance: all that we don't know is the value of its mean, which we estimate by the sample mean from a random sample of size N. The sample mean $\bar{X}$ is a *point* estimate of the population mean μ. However, we know from §10.2 that our single estimate should be interpreted as a random drawing from the population of all means that could be calculated from every sample of size N that could be drawn from the population: that is, it is a drawing from the distribution $\bar{X}\sim N\left(\mu,\sigma^2/N\right)$. Hence, as we have seen,

$$Z=\sqrt{N}\left(\frac{\bar{X}-\mu}{\sigma}\right)\sim N(0,1) \qquad (11.1)$$

Now suppose we are interested in the interval of values that will include Z with probability 0.95. This will be given by the probability statement

$$P\left(z_{0.975}<Z<z_{0.025}\right)=P\left(-z_{0.025}<Z<z_{0.025}\right)=0.95$$

where we use the symmetry of the normal distribution to equate $z_{0.975}$ with $-z_{0.025}$. Thus

$$P\left(-z_{0.025} < \sqrt{N}\left(\frac{\bar{X}-\mu}{\sigma}\right) < z_{0.025}\right) = P\left(-z_{0.025}\frac{\sigma}{\sqrt{N}} < \bar{X}-\mu < z_{0.025}\frac{\sigma}{\sqrt{N}}\right)$$

$$= P\left(-z_{0.025}\frac{\sigma}{\sqrt{N}} < \mu-\bar{X} < z_{0.025}\frac{\sigma}{\sqrt{N}}\right)$$

$$= P\left(\bar{X} - z_{0.025}\frac{\sigma}{\sqrt{N}} < \mu < \bar{X} + z_{0.025}\frac{\sigma}{\sqrt{N}}\right) = 0.95$$

where the expression after the second equality is obtained by multiplying the previous expression through by –1, and consequently changing the direction of the inequalities (for example, if $3 < x < 4$, then $-4 < x < -3$). The final expression

$$P\left(\bar{X} - z_{0.025}\frac{\sigma}{\sqrt{N}} < \mu < \bar{X} + z_{0.025}\frac{\sigma}{\sqrt{N}}\right) = 0.95$$

defines a 95% *confidence interval* for μ: the *upper bound* of the interval is

$$\mu_U = \bar{X} + z_{0.025}\frac{\sigma}{\sqrt{N}} = \bar{X} + 1.96\sigma_{\bar{X}}$$

and the *lower bound* is

$$\mu_L = \bar{X} - z_{0.025}\frac{\sigma}{\sqrt{N}} = \bar{X} - 1.96\sigma_{\bar{X}}$$

More generally, we can define a $100(1-\alpha)\%$ confidence interval (CI) for μ as

$$P\left(\bar{X} - z_{\alpha/2}\frac{\sigma}{\sqrt{N}} < \mu < \bar{X} + z_{\alpha/2}\frac{\sigma}{\sqrt{N}}\right) = 1-\alpha \tag{11.2}$$

Consider now the more realistic situation when we do not know the value of the population variance σ^2. However, we do know that an unbi ased and consistent estimator of σ^2 is given by s^2, so that replacing σ^2 by s^2 in (11.1) gives

$$\frac{\bar{X}-\mu}{s/\sqrt{N}} = \frac{\bar{X}-\mu}{\left(\frac{\sigma}{\sqrt{N}}\right)\sqrt{\frac{s^2}{\sigma^2}}} = \frac{Z}{\sqrt{\frac{s^2}{\sigma^2}}}$$

But, from §10.4,

$$\frac{s^2}{\sigma^2} \sim \frac{\chi^2(N-1)}{N-1}$$

so that

$$\sqrt{N}\frac{\bar{X}-\mu}{s} = \frac{Z}{\sqrt{\frac{s^2}{\sigma^2}}} \sim \frac{Z}{\sqrt{\frac{\chi^2(N-1)}{N-1}}}$$

From §9.4, this is the definition of a variable distributed as Student's t with N–1 degrees of freedom.[3] Thus

$$t = \sqrt{N}\frac{\bar{X}-\mu}{s} \sim t(N-1)$$

and, by analogous reasoning to that above, a $100(1-\alpha)\%$ CI for μ is

$$P\left(\bar{X} - t_{\alpha/2}\frac{s}{\sqrt{N}} < \mu < \bar{X} + t_{\alpha/2}\frac{s}{\sqrt{N}}\right) = 1-\alpha \qquad (11.3)$$

with *upper bound*

$$\mu_U = \bar{X} + t_{\alpha/2}\frac{s}{\sqrt{N}} = \bar{X} + t_{\alpha/2}s_{\bar{X}}$$

and *lower bound*

$$\mu_L = \bar{X} - t_{\alpha/2}\frac{s}{\sqrt{N}} = \bar{X} - t_{\alpha/2}s_{\bar{X}}$$

But what happens when the population distribution is not normal? If the sample is large then, via the CLT result in §9.3, we can continue to use the confidence interval given by (11.2). If the sample is small then, strictly, the interval given by (11.3) only applies under normality. However, as long as the departure from normality is not excessive, this interval based on the t distribution should continue to work reasonably well.

The interpretation of a confidence interval

Many users of statistical methods are notoriously confused by what confidence intervals actually mean! This is not surprising since the 'obvious' interpretation turns out to be wrong: it *does not* mean that there is, say, a 95% chance that the true mean lies within the interval. This is because the true mean, being a parameter of the population distribution, takes on one – and only one – value. It is thus *not a random variable* and hence either lies within the interval or does not: in this precise sense the statement that the true mean has a 95% chance of lying within any interval is meaningless.

A correct explanation of a 95% confidence interval is as follows: if we took many random samples of size N from a population with mean μ, we would find that μ lies within 95% of the calculated intervals. If, as is usually the case, we can only take one sample, then we do not (and cannot) know if the population mean will lie within our calculated interval, but we are comforted by the fact that the great majority of such intervals, on average 19 out of 20, will contain the population mean; we can thus be fairly sure – we say with confidence level 0.95 – that our single interval will indeed contain μ.

We illustrate this fundamental idea by an example using the income inequality data first introduced in §2.1. To reprise, this data set contains 189 observations on country incomes in 2009. Let us assume that these 189 observations represent the entire population of countries, although we know that this is not actually true (recall the discussion in §7.1). Let us further assume that the mean of this population is given by $\mu = 13389$ (the sample mean calculated using all 189 observations) and let us attempt to estimate this mean by taking random samples of size $N=20$ and constructing 95% confidence intervals using (11.3). These will use the critical t value $t_{0.025}(19)=2.093$, and the intervals so constructed from 20 such samples are shown in Figure 11.1.

Just one of the samples (coincidentally the first) does not contain the 'population' mean of 13389 (shown as the dashed horizontal line in Figure 11.1), so that 19 of the 20 do contain it, a fraction (fortuitously) exactly equal to the nominal *confidence level* of 0.95. Note that such a satisfactory result has been achieved even though we know that the 'population' distribution is certainly not normal, being highly skewed to the right. However, the widths of the confidence intervals vary quite widely. This is because we are estimating the unknown population variance by the sample variance s^2: across the 20 samples, the sample standard deviation s ranges from 7478 (sample 1) to 38298 (sample 19), so that the standard error $s_{\bar{X}}$ ranges from 1672 to 8564.

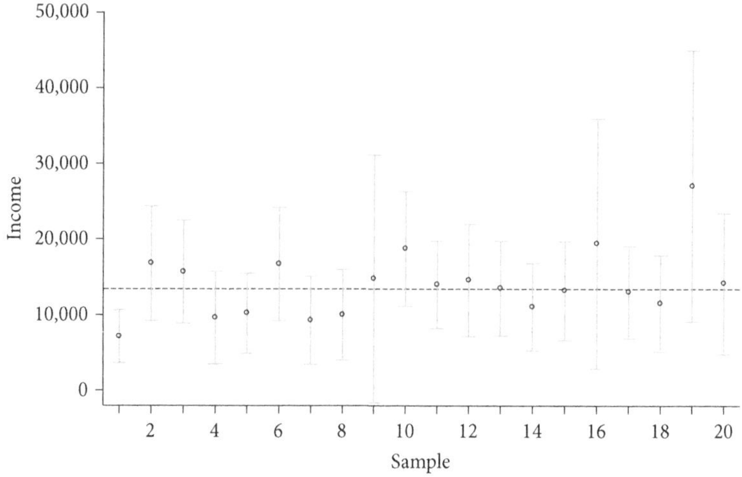

FIGURE 11.1 *Confidence intervals for 20 samples of size 20 drawn randomly from the income inequality data*

To illustrate the confidence interval calculations, sample 1 was found to have a mean of 7144, while sample 19 had a mean of 27072. The two 95% CIs were thus calculated (using the s_X values given above) as

Sample 1

$$7144 \pm 2.093\left(7478/\sqrt{20}\right) = 7144 \pm 3499 \quad \text{i.e.} \quad \left(3645, 10643\right)$$

Sample 19

$$27072 \pm 2.093\left(38298/\sqrt{20}\right) = 27072 \pm 17924 \quad \text{i.e.} \quad \left(9148, 44996\right)$$

11.3 A confidence interval for the population variance

The above example has shown that the sample standard deviation is rather variable across samples. Thus it would be useful to be able to construct a confidence interval for the population variance. From §10.4, we know that

$$\frac{(N-1)s^2}{\sigma^2} \sim \chi^2(N-1)$$

A $100(1-\alpha)\%$ CI for σ^2 is then obtained from

$$P\left(\chi^2_{1-\alpha/2}(N-1) \leq \frac{(N-1)s^2}{\sigma^2} \leq \chi^2_{\alpha/2}(N-1)\right) = 1-\alpha$$

as

$$P\left(\frac{(N-1)s^2}{\chi^2_{\alpha/2}(N-1)} \leq \sigma^2 \leq \frac{(N-1)s^2}{\chi^2_{1-\alpha/2}(N-1)}\right) = 1-\alpha$$

Thus, 95% confidence intervals for σ^2 calculated from samples 1 and 19, respectively, are

$$\frac{19 \times 7478^2}{\chi^2_{0.025}(19)} \leq \sigma^2 \leq \frac{19 \times 7478^2}{\chi^2_{0.975}(19)}$$

and

$$\frac{19 \times 38298^2}{\chi^2_{0.025}(19)} \leq \sigma^2 \leq \frac{19 \times 38298^2}{\chi^2_{0.975}(19)}$$

With $\chi^2_{0.975}(19) = 8.907$ and $\chi^2_{0.025}(19) = 32.85$, these intervals are, in terms of σ,

$$5687 \leq \sigma \leq 10922$$

and

$$29126 \leq \sigma \leq 55935$$

Since the standard deviation of the complete data set is 17629, we see that neither interval includes this value, although as we are calculating the confidence intervals using the two extreme estimates of s^2, this is perhaps not too surprising.

11.4 Hypothesis testing

So far, we have been considering how to estimate the unknown population parameters using sample data. Often, however, we wish to test *hypotheses* about the values taken by these parameters. Suppose that an economist, *before* looking at the income inequality data, claims that mean income is $\mu = 10000$. If a sample of just 20 countries was available, then one way to test the validity of this claim would be to construct, say, a 95% CI for the mean and see whether the value $\mu = 10000$ fell inside the calculated interval. If it did, we might conclude that the claim was justified; if it did not, we would be tempted to conclude that the claim was invalid. Looking again at the 20 confidence intervals shown in Figure 11.1, only one of the intervals (the tenth) does not contain $\mu = 10000$, so there is a good chance that we would, when taking just a single sample, conclude that the claim had some validity. Of course, we could be 'unlucky' and actually draw the tenth sample that does not contain $\mu = 10000$, in which case we would conclude, perhaps erroneously, that the claim was invalid.

Now suppose that a second economist claims that mean income is $\mu = 40000$. Now only one of the intervals contains this value, suggesting that the claim is likely to be false. However, it is again possible that the sample that is actually drawn turns out to be the 19th – the one that includes the value – in which case it would be concluded that the claim was valid. Further, suppose that, rather than computing 95% confidence intervals, we decided to compute 70% confidence intervals. Since $t_{0.015}(19) = 1.07$, such intervals stretch roughly one standard error on either side of the sample mean, and are thus around half the length of the 95% intervals. Now 11 of the 70% intervals will not contain $\mu = 10000$ (and certainly none will contain $\mu = 40000$), so that it is more likely that the claim will be found to be invalid. Thus decisions about claims are seen to be dependent upon both chance (that is, which sample is actually drawn) and on our choice of confidence level: there is thus no such thing as finding a claim *definitely* valid or invalid – each will have a probability of error attached to it.

These ideas can be formalised within the framework of a *hypothesis test*. A claim made by an economist, or indeed by anyone else, about the values taken by a population parameter is known as the *null hypothesis*, typically denoted H_0. The values not included in the claim become the *alternative hypothesis*, denoted here as H_A (although the notation H_1 is also commonly used). Typically, the null hypothesis will define a single value of the population parameter (say H_0: $\mu = 10000$), in which case the alternative will be all other values (the *two-sided alternative* H_A: $\mu \neq 10000$), although in some cases a one-sided alternative may be specified (say H_0: $\mu > 10000$ if it is felt that values of μ less than 10000 cannot possibly occur).

The validity of the null hypothesis is assessed by constructing a *test statistic*. This will depend upon the sample estimate of the population parameter and the value of the parameter under the null, and its exact form will depend upon the sampling distribution of the sample estimate. It will always be a random variable, however, and will thus follow a known probability distribution. A *rejection region*, based on a chosen *significance level* of the test, is then defined. If the calculated test statistic falls into this rejection region then the decision is to reject the null hypothesis; if the test statistic falls outside this rejection region then the decision is *not* to reject the null.[4]

Testing a hypothesis on the population mean

Suppose that we are interested in testing a hypothesis about the unknown mean μ of a normally distributed random variable. We have available a sample of size N containing the observations $X_1, X_2, \ldots, X_N$, from which we calculate the sample mean $\bar{X}$ and sample variance s^2. From §11.2, we know that

$$t = \sqrt{N}\,\frac{\bar{X} - \mu}{s} \sim t(N-1)$$

Suppose that the null hypothesis is that μ equals a particular value, say μ_0: formally, we state this as H_0: $\mu = \mu_0$, with the alternative hypothesis being H_A: $\mu \neq \mu_0$. We now *temporarily assume that H_0 is true*, in which case the test statistic

$$t_0 = \sqrt{N}\,\frac{\bar{X} - \mu_0}{s}$$

would be distributed as $t(N-1)$. Our next task is to choose the significance level of the test, α. This is often set at 5%, that is, $\alpha = 0.05$, although it could easily be chosen to be some other value. This choice defines the rejection region for the test. With our two-sided alternative, the rejection region is defined to be the tails of the distribution that, because of the symmetry of the t-distribution, contain $\alpha/2$% of the distribution in each of them, as shown pictorially in Figure 11.2.

Formally, the rejection region, RR, is defined as

$$RR: \left|t_0\right| \geq t_{\alpha/2}(N-1)$$

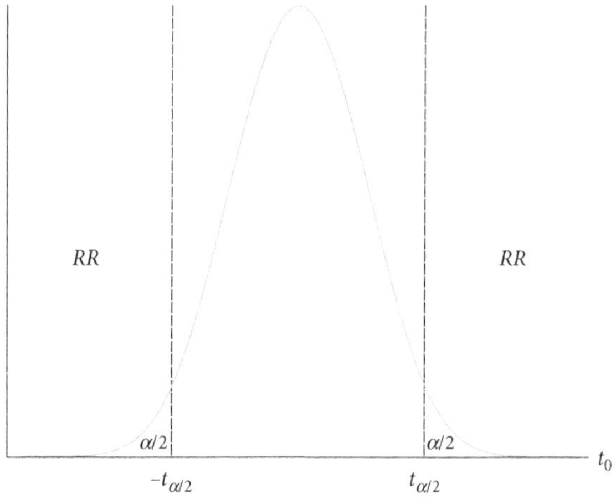

FIGURE 11.2 *A pictorial representation of a hypothesis test on a population mean*

The test can now be performed by calculating t_0 and establishing whether it falls in *RR* or not.

As an example, suppose that the null is H_0: $\mu = 10000$ and that, from a sample of $N = 20$, $\bar{X} = 18756$ and $s = 16248$ (that is, we have sample 10 from the income inequality data available). Then, assuming H_0 is true,

$$t_0 = \sqrt{20}\frac{18756 - 10000}{16248} = -2.41$$

The rejection region is defined, for $\alpha = 0.05$, as

$$RR: \left| t_0 \right| \geq t_{0.025}(19) = 2.093$$

Since $\left| t_0 \right| = 2.41 > 2.093$, the test statistic falls into *RR* and hence we reject the null hypothesis.

If the alternative is the one-sided H_A: $\mu > \mu_0$ then *all* of the rejection region falls into the right-hand tail of the distribution and so

$$RR: t_0 > t_\alpha (N-1)$$

with an analogous rejection region based on the left-hand tail if H_0: $\mu < \mu_0$.

The temporary assumption that the null hypothesis is true is a crucial feature of the test. By making this assumption, we ask the following question: what is the behaviour of the test statistic likely to be if the null is true? The answer is that t_0 will follow the appropriate t distribution, so that large (absolute) values of t_0 would be unlikely to be observed if the null was indeed true. If, nevertheless, we do observe such a large value, then this must cast doubt upon the appropriateness of assuming that the null hypothesis is true, and we consequently prefer to reject the null in favour of the alternative. Choosing the significance level α is thus a device that allows us to decide what we consider to be a sufficiently 'large' value of the test statistic to warrant rejection of the null.

Testing a hypothesis on the population variance

Suppose that we now wish to test a hypothesis concerning the unknown variance of a normal random variable. Here we can set up the null as $H_0\colon \sigma^2 = \sigma_0^2$ and the alternative as $H_A\colon \sigma^2 \neq \sigma_0^2$. In this case, again using the result that

$$\frac{(N-1)s^2}{\sigma^2} \sim \chi^2(N-1)$$

the test statistic is

$$\chi_0^2 = (N-1)\frac{s^2}{\sigma_0^2}$$

which will be distributed as $\chi^2(N-1)$ on the assumption that $H_0\colon\ \sigma^2 = \sigma_0^2$ is true. The rejection region for a test with α level of significance is defined as

$$RR: \chi_0^2 > \chi_{\alpha/2}^2(N-1) \quad \text{or} \quad \chi_0^2 < \chi_{1-\alpha/2}^2(N-1)$$

as can be seen from the pictorial representation in Figure 11.3.

As an example, suppose that $H_0\colon \sigma^2 - 15000^2$ and we again have sample 10. The test statistic is then

$$\chi_0^2 = 19\frac{16248^2}{15000^2} = 22.29$$

With $\alpha = 0.05$, $\chi_{0.975}^2(19) = 8.907$ and $\chi_{0.025}^2(19) = 32.85$, so that χ_0^2 does not fall in the rejection region and we cannot reject the null hypothesis.

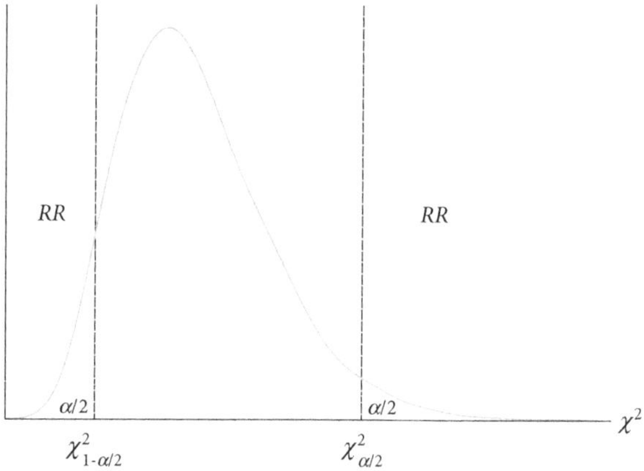

FIGURE 11.3 *A pictorial representation of a hypothesis test on a population variance*

11.5 Further considerations concerning hypothesis testing

The apparatus of hypothesis testing is crucial to all areas of economics, and several features of it are worth exploring further. The device of setting a significance level for a test has certain implications that data analysts need to be familiar with. The most important of these is to appreciate that, as has been made explicit in the development of hypothesis testing, incorrect decisions can, and often will, be made.

Consider again the setting up of a rejection region. The decision rule is to reject the null hypothesis if the test statistic falls in the rejection region. However, it is clear that the test statistic could fall into this region even if the null hypothesis is true: the choice of the significance level merely determines the probability of this happening. This is why the significance level is also known as the probability of making a *Type I Error*, formally defined as

$$\alpha = P\left(H_0 \text{ is rejected} \middle| H_0 \text{ is true}\right)$$

There is a second way that an incorrect decision may be made. This is when the null hypothesis is not rejected when the alternative hypothesis is true; it is known as the *Type II Error*. The probability of making a Type II error is known as β and defined as

$$\beta = P(H_0 \text{ is not rejected} | H_A \text{ is true})$$

It is straightforward to show that a trade-off exists between the two types of error: decreasing α necessarily implies that β increases. If the probability of incorrectly rejecting the null when it is true is made smaller, then the probability of incorrectly accepting the null when it is false must become larger.

Just as $1-\alpha$ is known as the *confidence level*, $1-\beta$ is called the *power* of the test: a good test is one that has, for a given confidence level, high power.

It is difficult to provide any firm guidelines for choosing the level of significance. Setting $\alpha = 0.05$ is traditional (and has the neat interpretation that a Type I error will only be made once every 20 tests on average), but it is by no means universally accepted or, indeed, always sensible. An alternative approach to hypothesis testing is to use *prob-values* (p-values). The p-value is the probability of observing a value of the test statistic at least as large as the one that has been calculated. Thus, when testing H_0: $\mu = 10000$, the test statistic $t_0 = 2.41$ was obtained. The probability of drawing such a value from a $t(19)$ distribution may be computed to be 0.0273, which states that there is a 2.73% chance of observing $|t_0| > 2.41$ if the null hypothesis was true: such a small value suggests that the null is very likely to be false and therefore should be rejected.

The use of p-values helps to avoid problems associated with the abrupt cut-offs implied by the rejection region approach. For example, suppose that, when testing H_0: $\mu = 10000$, a test statistic of 2.08 was obtained. This is below the 5% significance level of 2.093, but has a p-value of 0.0513, that is, it has a *marginal significance level* of 5.13%. There may be good grounds for rejecting this null even if the test statistic is not quite significant at the 5% level. One argument for doing so is that the size of the sample should be taken into consideration when significance levels are set. With very large sample sizes, standard errors of estimators become very small (the consistency argument), so that even small differences between the estimate and the null hypothesis will lead to significant test statistics, even though there may not be any 'economic' difference between them. On the other hand, with small samples standard errors are large, so that test statistics will be small and thus incapable of distinguishing between hypotheses. Thus there are good arguments for making significance levels inversely proportional to sample size: with small samples, set the significance level high, with large samples, set it small.[5]

Finally, confidence intervals and (two-sided) hypothesis tests are really 'two sides of the same coin'. For a given choice of α, if a test statistic falls into the rejection region then the value hypothesised under the null will lie outside the $100(1-\alpha)\%$ CI, so that in both cases a rejection of the null is indicated. Similarly, if the hypothesised value falls within the $100(1-\alpha)\%$ CI then the test statistic will not lie in the $\alpha\%$ rejection region, and the null will not be rejected.

11.6 Inference in correlation

Having developed a set of inferential procedures, we can now return to the analysis of correlation and begin to answer questions such as: What is the 95% CI for the correlation between consumption and income? and: Is the correlation between inflation and unemployment significant? To be able to do this, we require the sampling distribution of the *sample* correlation coefficient r_{XY}. This is made somewhat complicated by the fact that this distribution depends on the value of the unknown population correlation coefficient, which, on recalling the definition of covariance introduced in **§9.4**, we define as

$$\rho_{XY} = \frac{Cov(X,Y)}{\sqrt{V(X)V(Y)}} = \frac{E(XY)-E(X)E(Y)}{\sqrt{V(X)V(Y)}}$$

The sampling distributions of r for $\rho=0$ and $\rho=0.8$ are shown in Figure 11.4 (we drop the XY subscripts for notational simplicity).

To construct a CI for ρ we make use of the following transformation

$$f = \frac{1}{2}\ln\left(\frac{1+r}{1-r}\right) \sim N\left(\frac{1}{2}\ln\left(\frac{1+\rho}{1-\rho}\right), \frac{1}{N-3}\right)$$

where r is computed from the sample $(X_1,Y_1),\ldots,(X_N,Y_N)$. Thus, a $100(1-\alpha)\%$ CI for ρ is obtained by first calculating the interval

$$f \pm z_{\alpha/2}\frac{1}{\sqrt{N-3}}$$

to give upper and lower bounds f_U and f_L. We then need to solve

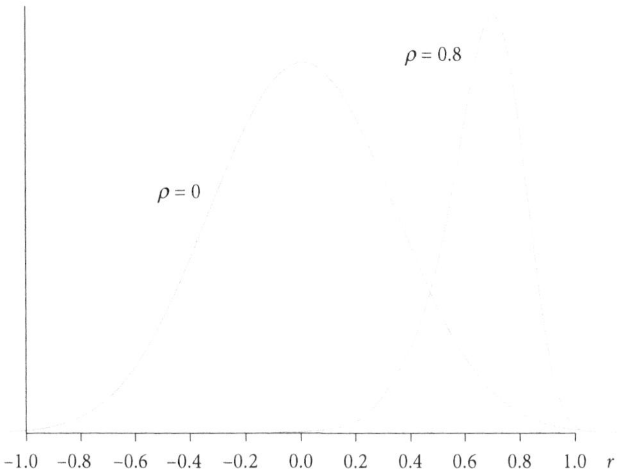

FIGURE 11.4 *Distribution of the population correlation coefficient ρ*

$$f = \frac{1}{2}\ln\left(\frac{1+\rho}{1-\rho}\right)$$

in terms of ρ. Some algebra shows this to be

$$\rho = \frac{\exp(2f)-1}{\exp(2f)+1}$$

and this formula can then be used to calculate ρ_U and ρ_L. For example, in §5.1 we found that the sample correlation between U.K. consumption and income was $r=0.997$ when calculated using $N=63$ post-war annual observations.

A 95% confidence interval for the population correlation is thus calculated in two steps. First, on noting that

$$f = \frac{1}{2}\ln\left(\frac{1+0.997}{1-0.997}\right) = 3.2504$$

we calculate the interval

$$3.2504 \pm 1.96\frac{1}{\sqrt{63}} = 3.2504 \pm 0.2469$$

yielding $f_U = 3.4973$ and $f_L = 3.0035$.

Transforming back to correlations, we obtain

$$\rho_U = \frac{\exp(2 \times 3.4973) - 1}{\exp(2 \times 3.4973) + 1} = 0.998$$

and

$$\rho_L = \frac{\exp(2 \times 3.0035) - 1}{\exp(2 \times 3.0035) + 1} = 0.995$$

so that a 95% confidence interval is $0.995 < \rho < 0.998$, which, as expected, is very 'tight', but asymmetric about $r = 0.997$.

This transformation, known as the *Fisher transformation*, can easily be used to construct a test of the hypothesis H_0: $\rho = \rho_0 \neq 0$, but such tests are rarely used or of interest in economics. Of much more importance is a test of the null hypothesis H_0: $\rho = 0$, as this states that X and Y are uncorrelated in the population. Here we can make use of the result, alluded to above, that r is approximately t distributed under this null. In fact, the test statistic that we use is

$$t = r\sqrt{\frac{N-2}{1-r^2}} \sim t(N-2)$$

In §5.1 we also found that, using $N = 156$ annual observations, the correlation between U.K. inflation and unemployment was −0.34. The above test statistic is

$$t = -0.34\sqrt{\frac{156}{1-0.34^2}} = -4.52 \sim t(154)$$

which has a p-value less than 0.00002 and is thus clearly significant: we may thus reject the hypothesis that inflation and unemployment are uncorrelated.

This approach to testing the null of zero correlation may easily be extended to partial correlations. In this situation, where we are holding a third variable 'constant', the degrees of freedom are reduced by one: that is, N–3 appears in both the test statistic and the degrees of freedom. For example, the partial correlation between salary and education with experience held constant was $r_{XY.Z} = 0.414$, calculated using $N = 12$ observations. To test H_0: $\rho_{XY.Z} = 0$, we calculate

$$t = 0.414\sqrt{\frac{9}{1-0.414^2}} = 1.36 \sim t(9)$$

This has a *p*-value of 0.21 and is clearly insignificant.[6]

Notes

1 A proof of this result is extremely complicated, and was first shown by Karl Pearson, 'On the standard error of the median ...', *Biometrika* 23 (1931), 361–363.

2 A proof of this result, which is based on the *Cramer-Rao inequality* (or *lower bound*), is given in Paul G. Hoel, *Introduction to Mathematical Statistics*, 4th edition (Wiley, 1971), p. 365.

3 This result assumes that the numerator and denominator are independent of each other, and it can be shown that this is the case here.

4 Hypothesis testing plays a central role in statistical inference and its compatriot, *statistical significance*, has been the subject of many philosophical and methodological debates since the development of the competing inferential frameworks of Sir Ronald Fisher and Jerzy Neyman and Egon Pearson in the 1920s and 1930s: see, for example, Johannes Lenhard, 'Models and statistical inference: the controversy between Fisher and Neyman-Pearson', *British Journal of the Philosophy of Science* 57 (2006), 69–91. This debate has flared up again recently in economics: see Stephen T. Ziliak and Deirdre N. McCloskey, *The Cult of Statistical Significance: How the Standard Error Costs Us Jobs, Justice and Lives* (University of Michigan Press, 2008) and Kevin D. Hoover and Mark V. Siegler, 'Sound and fury: McCloskey and significance testing in economics', *Journal of Economic Methodology* 15 (2008), 1–37. A framework that may have the potential of reconciling the various approaches to hypothesis testing is the *severe testing* methodology of Deborah G. Mayo and Aris Spanos, 'Severe testing as a basic concept in a Neyman–Pearson philosophy of induction', *British Journal of the Philosophy of Science* 57 (2006), 323–357. For more economic-centred discussion of this idea, see John DiNardo, 'Interesting questions in *Freakonomics*', *Journal of Economic Perspectives* 45 (2007), 973–1000, and Terence C. Mills, 'Severe hypothesis testing in economics', *Journal of Quantitative Economics* 7 (2009), 1–19.

5 Such an argument is often regarded as being an implication of Bayesian inference (recall §7.2), but it is also consistent with the severe testing approach.

6 For the rank correlation r^s introduced in §5.3, the null H_0: $\rho^s = 0$ can be tested using the statistic $z^s = r^s\sqrt{N-1} \sim N(0,1)$.

12
The Classical Linear Regression Model

Abstract: *The regression model of Chapter 6 is revisited using the inferential framework developed in subsequent chapters. The theory underlying the least squares approach is developed in more detail, so providing the 'algebra' of regression. The concepts of population and sample regression functions are introduced, along with the 'classical assumptions' of regression. These assumptions allow the ordinary least squares (OLS) estimators to satisfy the Gauss–Markov theorem, thus becoming best linear unbiased estimators, this being illustrated by simulation. Statistical inference in regression is then developed along with a geometrical interpretation of hypothesis testing. Finally, the use of regressions for prediction and considerations of functional form and non-linearity are discussed. Several examples are used to illustrate these concepts throughout the chapter.*

12.1 The algebra of regression

In §6.2 we introduced (bivariate) regression analysis, in which a sample of N pairs of observations on the variables X and Y, $(X_1,Y_1), (X_2,Y_2), \ldots, (X_N, Y_N)$ were used to fit the regression line

$$\hat{Y}_i = \hat{\beta}_0 + \hat{\beta}_1 X_i$$

by the method of least squares.[1] As we saw there, mathematically the *least squares regression problem* is to choose those values of $\hat{\beta}_0$ and $\hat{\beta}_1$ that minimise the sum of squared residuals

$$S = \sum_{i=1}^{N} \hat{u}_i^2 = \sum_{i=1}^{N} \left(Y_i - \hat{\beta}_0 - \hat{\beta}_1 X_i\right)^2$$

Thus S has to be differentiated with respect to $\hat{\beta}_0$ and $\hat{\beta}_1$ and the derivatives set to zero (subscripts and summation limits are now omitted for clarity):

$$\frac{\partial S}{\partial \hat{\beta}_0} = 2\sum\left(Y - \hat{\beta}_0 - \hat{\beta}_1 X\right)(-1) = 0$$

$$\frac{\partial S}{\partial \hat{\beta}_1} = 2\sum\left(Y - \hat{\beta}_0 - \hat{\beta}_1 X\right)(-X) = 0$$

i.e.,

$$\sum\left(Y - \hat{\beta}_0 - \hat{\beta}_1 X\right) = 0$$

$$\sum\left(YX - \hat{\beta}_0 X - \hat{\beta}_1 X^2\right) = 0$$

These 'first order conditions' can be written as the *normal equations*

$$\sum Y - N\hat{\beta}_0 - \hat{\beta}_1 \sum X = 0 \tag{12.1}$$

and

$$\sum YX - \hat{\beta}_0 \sum X - \hat{\beta}_1 \sum X^2 = 0 \tag{12.2}$$

The first of these, (12.1), can be solved for $\hat{\beta}_0$

$$\hat{\beta}_0 = \frac{\sum Y}{N} - \hat{\beta}_1 \frac{\sum X}{N} = \bar{Y} - \hat{\beta}_1 \bar{X} \tag{12.3}$$

Then substituting (12.3) into (12.2),

$$\sum XY - \left(\frac{\sum Y}{N} - \hat{\beta}_1 \frac{\sum X}{N} \right) \sum X - \hat{\beta}_1 \sum X^2 = 0$$

and rearranging gives

$$\left(\left(\sum X^2\right) - \frac{\left(\sum X\right)^2}{N} \right) \hat{\beta}_1 = \sum XY - \frac{\sum X \sum Y}{N}$$

which can then be solved for $\hat{\beta}_1$:

$$\hat{\beta}_1 = \frac{N\sum XY - \sum X \sum Y}{N\sum X^2 - \left(\sum X\right)^2} = \frac{\sum xy}{\sum x^2} \tag{12.4}$$

where the second equality uses the equivalences derived in note 2 of Chapter 4.

Strictly, setting the first derivatives $\partial S / \partial \hat{\beta}_0$ and $\partial S / \partial \hat{\beta}_1$ to zero for the first order conditions only ensures an extremum solution, which could correspond to either a minimum or a maximum sum of squared residuals. However, the second order conditions for a minimum, which involve

$$\frac{\partial^2 S}{\partial \hat{\beta}_0^{\,2}} = 2N \qquad \frac{\partial^2 S}{\partial \hat{\beta}_1^{\,2}} = 2\sum X^2 \qquad \frac{\partial^2 S}{\partial \hat{\beta}_0 \partial \hat{\beta}_1} = 2\sum X$$

are that $\partial^2 S / \partial \hat{\beta}_0^2 > 0$ and $\partial^2 S / \partial \hat{\beta}_1^2 > 0$, which are both clearly satisfied, and

$$\frac{\partial^2 S}{\partial \beta_0^2} \frac{\partial^2 S}{\partial \beta_1^2} - \frac{\partial^2 S}{\partial \beta_0 \partial \beta_1} > 0$$

This implies that $\sum X^2 - \left(\sum X\right)^2 / N = \sum \left(X - \bar{X}\right)^2 > 0$, which is clearly the case, so the second order conditions do indeed ensure that the normal equations yield a minimum.

As the residuals are given by

$$\hat{u}_i = Y_i - \hat{\beta}_0 - \hat{\beta}_1 X_i$$

the normal equations (12.1) and (12.2) can be written as

$$\sum \hat{u} = 0$$
$$\sum x\hat{u} = 0$$

that is, the least squares fit is such that the sum of the residuals is zero, as is the sum of the product of the residuals and the regressor *X*, measured in deviations form.

The residual sum of squares (*S*; now defined as *RSS*) is (cf. a similar derivation in §6.4):

$$\begin{aligned}
RSS &= \sum \hat{u}^2 = \sum\left(Y - \hat{\beta}_0 - \hat{\beta}_1 X\right)^2 \\
&= \sum\left(Y - \bar{Y} + \hat{\beta}_1\bar{X} - \hat{\beta}_1 X\right)^2 \\
&= \sum\left(\left(Y - \bar{Y}\right) - \hat{\beta}_1\left(X - \bar{X}\right)\right)^2 \\
&= \sum\left(y - \hat{\beta}_1 x\right)^2 \\
&= \sum\left(y^2 + \hat{\beta}_1^2 x^2 - 2\hat{\beta}_1 xy\right) \\
&= \sum y^2 + \hat{\beta}_1^2 \sum x^2 - 2\hat{\beta}_1 \sum xy \\
&= \sum y^2 - \hat{\beta}_1 \sum xy
\end{aligned}$$

using $\hat{\beta}_1 = \sum xy / \sum x^2$. If we denote $\sum y^2$ as the *Total Sum of Squares* (*TSS*) and $\hat{\beta}_1 \sum xy$ as the *Explained Sum of Squares* (*ESS*), then we have the 'decomposition'

$$RSS = TSS - ESS$$

We can now define the *Coefficient of Determination*, which measures goodness of fit and was first introduced in §6.4, as

$$\begin{aligned}
r_{XY}^2 &= \frac{ESS}{TSS} = \frac{TSS - RSS}{TSS} = 1 - \frac{RSS}{TSS} \\
&= \frac{\hat{b}_1 \sum xy}{\sum y^2} = \frac{\left(\sum xy\right)^2}{\sum x^2 \sum y^2} = \hat{b}_1^2 \frac{\sum x^2}{\sum y^2}
\end{aligned}$$

From the formula (12.3) for $\hat{\beta}_0$, we have

$$\bar{Y} = \hat{\beta}_0 + \hat{\beta}_1 \bar{X}$$

that is, the fitted line passes through the point of the sample means $(\bar{Y}, \bar{X})$: this, of course, *does not* need to be an observation point. Furthermore, since $\hat{u}_i = Y_i - \hat{Y}_i$, so that

$$Y_i = \hat{\beta}_0 + \hat{\beta}_1 X_i + \hat{u}_i$$

we then have

$$Y_i - \bar{Y} = \hat{\beta}_1 (X_i - \bar{X}) + \hat{u}_i$$

or

$$y_i = \hat{\beta}_1 x_i + \hat{u}_i$$

and

$$\hat{y}_i = \hat{\beta}_1 x_i$$

(cf. equation (6.7)). This implies that, if we work in *deviations about means*, we can ignore the intercept in our derivations: geometrically, we have *shifted the axes* from origin (0,0) to $(\bar{Y}, \bar{X})$.

Fitting the UK consumption function

In §6.7 an estimated log–linear consumption function for the UK for the period 1948–2010 was reported as $\ln\hat{C} = 0.09 + 0.98 \ln Y$. The calculations underlying this regression are, with c and y representing deviations of the logarithms of consumption and income from their respective means, $\overline{\ln C}$ and $\overline{\ln Y}$:

$$\bar{C} = 4.97 \qquad \overline{\ln Y} = 13.4488$$

$$\sum c^2 = 12.2854 \qquad \sum y^2 = 12.7003 \qquad \sum cy = 12.4705$$

Thus

$$\hat{\beta}_1 = \frac{\sum cy}{\sum y^2} = \frac{12.4705}{12.7003} = 0.982$$

$$\hat{\beta}_0 = \overline{\ln C} - \hat{\beta}_1 \overline{\ln Y} = 13.2926 - (0.982 \times 13.4488) = 0.087$$

For this regression

$$TSS = 12.2854$$

$$ESS = 0.982 \times 12.4705 = 12.2451$$

$$RSS = TSS - ESS = 0.0403$$

so that the goodness of fit is

$$r_{YC}^2 = \frac{12.2451}{12.2854} = 1 - \frac{0.0403}{12.2854} = 0.9967$$

so that variation in the logarithms of income explains 99.7% of the variation in the logarithms of consumption.

12.2 Population and sample regression functions

Having developed the mechanics of linear regression, we need to provide some theoretical underpinnings to the technique so that we can assess under what circumstances it can, and cannot, be used to analyse economic data. To do this, we introduce the concept of the *population regression function* (PRF), which we may regard as a theoretical relationship between Y and X of the form

$$E(Y_i | X_i) = \beta_0 + \beta_1 X_i$$

where $E(Y_i | X_i)$ is the mean of Y *conditional* on X.

Equivalently, we may write

$$\begin{aligned} Y_i &= \beta_0 + \beta_1 X_i + u_i \\ &= E(Y_i | X_i) + u_i \end{aligned}$$

where

$$u_i = Y_i - E(Y_i | X_i)$$

The error u_i is the population counterpart of the residual, and captures the effect of all other influences on Y, apart from X, that we either choose not to model or which cannot be modelled.

Following the approach in previous chapters, we can also define the *sample regression function* (SRF), which is simply our fitted regression line

$$\hat{Y}_i = \hat{\beta}_0 + \hat{\beta}_1 X_i$$

We know that the SRF is the 'line of best fit' using a least squares criterion – but do the *estimators* $\hat{\beta}_0$ and $\hat{\beta}$ have any desirable properties in themselves? In fact they do, if we make a certain set of assumptions for the PRF, known as the *classical assumptions*.

1. $E(u_i) = 0$ for all i.

This assumes that the errors have zero mean.

2. $V(u_i) = \sigma^2$ for all i.

This assumes that the errors have a common variance, a property known as *homoskedasticity*. Note that the combination of assumptions 1 and 2 implies $E\left(u_i^2\right) = \sigma^2$, since $V(u_i) = E(u_i - E(u_i))^2 = E(u_i^2 - 0) = E(u_i^2)$.

3. $Cov(u_i, u_j) = 0$ for all $i \neq j$.

This assumes that any two errors have zero covariance and hence are uncorrelated. Since $Cov(u_i, u_j) = E(u_i u_j) - E(u_i)E(u_j)$, this assumption can equivalently be written as $E(u_i u_j) = 0$.

4. $Cov(u_i, u_j) = 0$ for all i and j.

This assumes either that the errors and the regressor are independent and hence have zero covariance or that the x_j are non-stochastic. It can equivalently be written as $E(u_i x_j) = 0$, which in the latter case becomes $E(u_i x_j) = x_j E(u_i) = 0$.[2]

5. The u_i are normally distributed for all i.

With assumptions 1–3, this implies that the errors are normally and independently distributed with mean zero and common variance σ^2, which we denote as $u_i \sim IN(0, \sigma^2)$.

12.3 Properties of the ordinary least squares estimators

The estimators defined in §12.1 as (12.3) and (12.4) are known as the ordinary least squares (OLS) estimators:

$$\hat{\beta}_1 = \frac{\sum xy}{\sum x^2} \qquad \hat{\beta}_0 = \bar{Y} - \hat{\beta}_1 \bar{X}$$

With assumptions 1 to 4, these OLS estimators satisfy the *Gauss–Markov theorem*:

> Within the class of *unbiased linear* estimators, the OLS estimators have *minimum variance*, that is, they are the *best linear unbiased estimators* (BLUE).

What does this theorem, which lies at the heart of regression analysis and hence of econometrics, actually mean? The following points are crucial in understanding its importance.

(i) We restrict attention only to *linear* estimators. This means that the formula for an estimator must be a linear function (in fact, a linear combination) of the y_is:

$$\hat{\beta}_1 = \frac{\sum x_i y_i}{\sum x_i^2} = w_1 y_1 + w_2 y_2 + \ldots + w_N y_N = \sum w_i y_i$$

where the weights are given by

$$w_i = \frac{x_i}{\sum x_i^2}$$

(ii) Given that we are only considering linear estimators, we further restrict attention to *unbiased* estimators: those estimators that, *on average*, yield the true value of the parameter they are estimating. Formally, as we saw in §**11.1**, an unbiased estimator of β_1, for example, has the property that $E\left(\hat{\beta}_1\right) = \beta_1$. That the OLS estimator is unbiased is easily shown:

$$E\left(\hat{\beta}_1\right) = E\left(\sum wy\right) = E\left(\sum w\beta_1 x\right) = \beta_1 E\left(\sum wx\right) = \beta_1 E(1) = \beta_1$$

on noting that $\sum wx = \sum\left(x^2 / \sum x^2\right) = \sum x^2 / \sum x^2 = 1$.

How do we interpret an unbiased estimator? Let us perform the following experiment, known as a *Monte Carlo simulation*. Suppose that we are in the unrealistic position of actually knowing the PRF between Y and a variable X, which takes the values

$$X_1 = 1,\ X_2 = 2,\ \ldots,\ X_{100} = 100$$

that is, X is a time trend: $X_t = t$ with $T = 100$. This known PRF is $Y_t = 9 + t + u_t$, where $u_t \sim IN(0,10)$, so that $Y_1 = 10 + u_1$, $Y_2 = 11 + u_2$, etc. Note that the *true* value of the slope is $\beta_1 = 1$.Now suppose that we

(a) Draw a large number, say 1,000, of samples of size 100 from our u distribution. This is simple to do with random number generators in modern econometric computer packages. Note that these samples will all be different.
(b) For each u sample, calculate $Y_t = 9 + t + u_t$. This will give us 1,000 samples of size 100 of (Y_t, t) data.
(c) Run the 1,000 regressions of Y on t, calculating $\hat{\beta}_1$ each time.
(d) Construct the frequency distribution of $\hat{\beta}_1$ (typically represented as a histogram: recall §**2.1**): this should be centred on $\beta_1 = 1$ if $\hat{\beta}_1$ is unbiased.

The histogram so constructed is shown in Figure 12.1(a), from which the unbiasedness property is clearly seen.

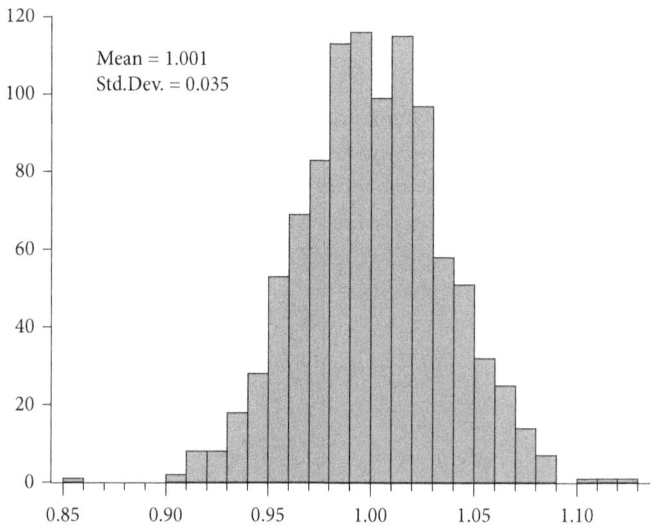

FIGURE 12.1(a) *Monte Carlo distributions of two alternative estimators of* β_1; *distribution of* $\hat{\beta}_1$

Furthermore, the distribution looks normal: in this experiment this is a consequence of u being normal, but it would in fact hold via the CLT for any distribution of u as long as the sample was of a reasonable size (recall §10.3).

(iii) Note that, because the β_1 differ across the samples, we obtain a frequency distribution with, obviously, a variance associated with it. The *best* (minimum variance) property of the OLS estimator states that this variance is the smallest possible within the class of linear unbiased estimators.

To illustrate this property, consider the following alternative estimator of β_1.

$$\tilde{\beta}_1 = \frac{Y_{100} - Y_1}{X_{100} - X_1} = \frac{Y_{100} - Y_1}{99}$$

$\tilde{\beta}_1$ is linear and can be shown to be unbiased, but uses only two observations, ignoring all the rest! Not surprisingly, the variance of its frequency distribution calculated from the same 1,000 samples is far greater than that for β_1, being approximately 18 times as large (see Figure 12.1(b)).

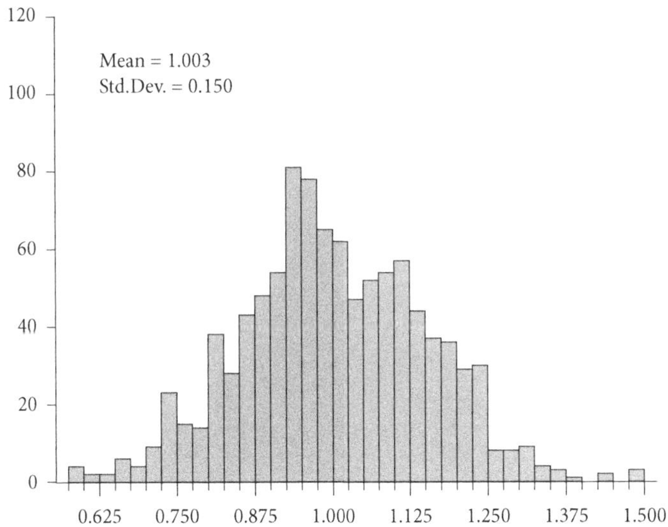

FIGURE 12.1(b) *Monte Carlo distributions of two alternative estimators of β_1; distribution of $\tilde{\beta}_1$*

As we can see from the two histograms, this means that there is a far greater chance that we will (unluckily) obtain an estimate of β_1 that is a considerable distance from the true value of 1, and hence obtain an inaccurate estimate of the slope.

More importantly, it will also be the case that *all* other linear unbiased estimators will have a greater variance than the OLS estimator (these remarks apply equally to estimators of the intercept, β_1). Thus the OLS estimator is relatively efficient and is said to be *best* in the class of unbiased linear estimators.[3]

12.4 Statistical inference in regression

Given the five assumptions made in §12.2, and now knowing that OLS is BLUE, we can state the following results about $\hat{\beta}_0$ and $\hat{\beta}_1$:[4]

$$E\left(\hat{\beta}_0\right)=\beta_0 \qquad E\left(\hat{\beta}_1\right)=\beta_1$$

$$V\left(\hat{\beta}_0\right)=\sigma^2\left(\frac{1}{N}+\frac{\bar{X}^2}{\sum x^2}\right) \qquad V\left(\hat{\beta}_1\right)=\frac{\sigma^2}{\sum x^2}$$

$$Cov\left(\hat{\beta}_0,\hat{\beta}_1\right)=\sigma^2\left(\frac{-\bar{X}}{\sum x^2}\right)$$

Furthermore, $\hat{\beta}_0$ and $\hat{\beta}_1$ are jointly normally distributed. Note that the formulae for the variances and covariance contain σ^2, which will typically be unknown. An unbiased estimator of the error variance is

$$\hat{\sigma}^2=\frac{RSS}{N-2}=\frac{\sum \hat{u}^2}{N-2}$$

With this we can define the standard errors of the estimators as

$$SE\left(\hat{\beta}_0\right)=\hat{\sigma}\sqrt{\left(\frac{1}{N}+\frac{\bar{X}^2}{\sum x^2}\right)}$$

$$SE\left(\hat{\beta}_1\right)=\frac{\hat{\sigma}}{\sqrt{\sum x^2}}$$

Now consider the variable

$$t_1 = \frac{\hat{\beta}_1 - \beta_1}{SE(\hat{\beta}_1)} = \frac{\hat{\beta}_1 - \beta_1}{\dfrac{\hat{\sigma}}{\sqrt{\sum x^2}}}$$

This can be written as

$$t_1 = \frac{\hat{\beta}_1 - \beta_1}{\sqrt{\dfrac{\sigma^2}{\sum x^2}}\sqrt{\dfrac{\hat{\sigma}^2}{\sigma^2}}} = \frac{\hat{\beta}_1 - \beta_1}{V\left(\hat{\beta}_1\right)\sqrt{\dfrac{\hat{\sigma}^2}{\sigma^2}}}$$

Since $\left(\hat{\beta}_1 - \beta_1\right)\Big/\sqrt{V\left(\hat{\beta}_1\right)} \sim N(0,1)$ and, by an obvious extension of the result in §10.4, $\hat{\sigma}^2/\sigma^2 \sim \chi^2(N-2)/(N-2)$, it therefore follows that $t_1 \sim t(N-2)$.[5]

Suppose we wish to test the null hypothesis

$$H_0 : \beta_1 = \beta_1^*$$

against the two-sided alternative $H_A : \beta_1 \neq \beta_1^*$. If the null hypothesis is true, then

$$t_1 = \frac{\hat{\beta}_1 - \beta_1^*}{SE\left(\hat{\beta}_1\right)} \sim t(N-2)$$

and hence the null will be rejected if t_1 is sufficiently large in absolute value. If we test at the α significance level, then our decision rule is to

$$\text{reject } H_0 \text{ if } |t_1| > t_{\alpha/2}(N-2)$$

where $t_{\alpha/2}(N-2)$ is the $\alpha/2$ percentage point of the $t(N-2)$ distribution. If the alternative is one-sided, say $H_A : \beta_1 > \beta_1^*$, then our decision rule is to

$$\text{reject } H_0 \text{ if } t_1 > t_\alpha(N-2)$$

Salaries, education and experience revisited

Let us consider again the salaries–education regression example of §6.2. Recall that we obtained

$$\hat{\beta}_0 = 15.263 \qquad \hat{\beta}_1 = 2.947$$

The standard errors of these estimates are calculated as:

$$SE\left(\hat{\beta}_0\right) = \hat{\sigma}\sqrt{\frac{1}{12} + \frac{5^2}{38}} = 0.861\hat{\sigma}$$

$$SE\left(\hat{\beta}_1\right) = \frac{\hat{\sigma}}{\sqrt{38}} = \frac{\hat{\sigma}}{6.164}$$

Now

$$\hat{\sigma} = \sqrt{\frac{538 - (2.947 \times 112)}{10}} = 4.56$$

so that

$$SE\left(\hat{\beta}_0\right) = 3.926$$

$$SE\left(\hat{\beta}_1\right) = 0.740$$

Suppose we wish to test the hypothesis that salary in the absence of any post-school education is £10,000. Thus to test H_0: $\beta_0 = 10$ against the two-sided alternative H_A: $\beta_0 \neq 10$, we compute

$$t_0 = \frac{15.263 - 10}{3.926} = 1.34$$

If we use an $\alpha = 0.05$ significance level, so that $t_{0.025}(10) = 2.23$, then since $|1.34| < 2.23$, we do not reject the null hypothesis.

We may wish to test the null hypothesis that post-school education has no effect on salary against the alternative hypothesis that it has a positive effect. This is a test of H_0: $\beta_1 = 0$ against the one-sided alternative H_A: $\beta_1 > 0$. Using $\alpha = 0.05$, we compute

$$t_1 = \frac{2.947 - 0}{0.740} = 3.98$$

Since $t_{0.05}(10) = 1.81 < t_1$, we can reject the null hypothesis. Tests of a coefficient null hypothesis using a t distribution are usually known as 't-tests', with the test statistic being referred to as a 't-statistic'. When the null has a value of zero, the latter is typically known as a 't-ratio', for obvious reasons.

Inference in the consumption function

A conventional format for reporting regression results is

$$\begin{array}{lll} \hat{Y}_i = \hat{\beta}_0 & + \quad \hat{\beta}_1 X_i & \\ \quad (SE) & \quad (SE) & R^2 \quad \hat{\sigma} \\ [t-ratio] & [t-ratio] & \end{array}$$

where we now write R^2 for r_{XY}^2 for consistency with subsequent notation. Without going into computational details, the linear and logarithmic consumption functions are estimated to be

$$\begin{array}{lll} \hat{C}_t = -4708 + 0.860\, Y_t & & \\ \quad (4328) \quad (0.005) & R^2 = 0.9978 & \hat{\sigma} = 13845 \\ \quad [1.09] \quad [166.1] & & \end{array}$$

$$\begin{array}{lll} \ln \hat{C}_t = -0.087 + 0.982 \ln Y_t & & \\ \quad (0.097) \quad (0.007) & R^2 = 0.9967 & \hat{\sigma} = 0.0257 \\ \quad [0.90] \quad [136.2] & & \end{array}$$

Using appropriate critical values of the $t(61)$ distribution, it can be ascertained that the intercepts in both functions are insignificantly different from zero but that the slopes are significantly less than one.

It is often tempting to compare regression models on the basis of goodness of fit, by for example, selecting the regression with the highest R^2 or lowest $\hat{\sigma}$. This is a dangerous procedure in any event, but it is *incorrect* to compare these two regressions on the basis of R^2 and $\hat{\sigma}$ values as the dependent variables are *different* (a method by which they may be compared is suggested in §15.6). Note also that it is not good practice to report estimates to many decimal places, as this conveys a sense of 'false precision' and obfuscates interpretation.

Confidence intervals

A $100(1-\alpha)\%$CI for β_1 is

$$\hat{\beta}_1 \pm t_{\alpha/2}(N-2)SE\left(\hat{\beta}_1\right)$$

With $\alpha = 0.05$, a 95% CI for β_1 from our salary–education regression is

$$2.947 \pm (2.23)(0.740)$$

i.e.

$$2.947 \pm 1.650$$

or

$$P\left[1.297 < \beta_1 < 4.597\right] = 0.95$$

The interval does not contain zero, thus confirming that β_1 is significantly positive. Similarly, for β_0 we have

$$15.263 \pm (2.23)(3.926)$$

i.e.

$$15.263 \pm 8.755$$

or

$$P\left[6.508 < \beta_0 < 24.018\right] = 0.95$$

While this interval does not contain zero, so that β_0 is certainly significant, the width of the interval nevertheless shows that the intercept is quite imprecisely estimated.

12.5 A geometric interpretation of hypothesis testing

Suppose we have fitted a regression to the scatterplot of data shown below and we are then interested in testing the hypothesis H_0: $\beta_1 = 0$, i.e., that Y does not depend on X. We can interpret this null hypothesis as the horizontal line with intercept $\bar{Y}$. The OLS regression line $\hat{Y} = \hat{\beta}_0 + \hat{\beta}_1 X$ produces a residual sum of squares, which we now denote RSS_U, which cannot be made any smaller.

The hypothesis H_0: $\beta_1 = 0$ imposes a restriction on the slope which forces the line to pivot about $\bar{X}$ and become horizontal (recall that *every* regression line passes through the point of the sample means). This horizontal line will produce a residual sum of squares, RSS_R, such that $RSS_R - RSS_U > 0$.

The crucial question is: by *how much is* RSS_R greater than RSS_U? If the increase in the residual sum of squares is large, then the restricted (horizontal) line provides a poor fit and H_0: $\beta_1 = 0$ should be rejected. If the increase is only small, then the restricted line might be judged to provide a satisfactory fit, and if this is the case H_0: $\beta_1 = 0$ should not be rejected.

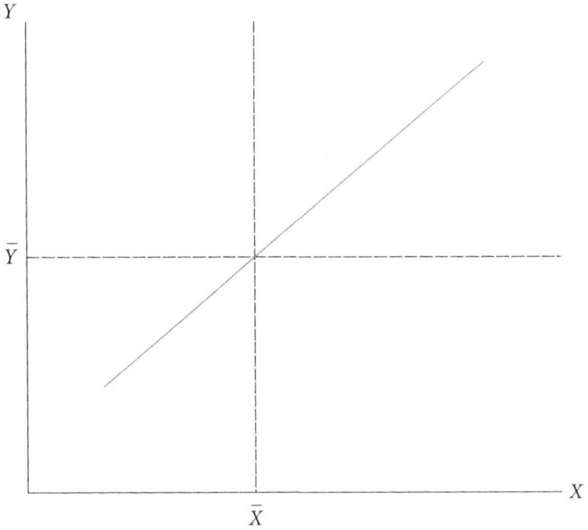

How, though, do we judge whether the increase in the residual sum of squares is sufficiently large to warrant the rejection of H_0: $\beta_1 = 0$, or whether the increase is so small as to be simply explained by sampling errors, so leading us to conclude that the null is, in fact, 'true'? We can make such a judgement using the statistic

$$F = \frac{(RSS_R - RSS_U)/2}{RSS_U/(N-2)} \sim F(1, N-2) \tag{12.5}$$

so that if $F > F_\alpha(1, N-2)$ we reject H_0: $\beta_1 = 0$ at the α significance level. The formula for F, with amendments for the two sets of degrees of freedom, is a quite general formulation, but in this case, since $Y = \beta_0 + u$ under the null, estimated to be $Y = \bar{Y} + \hat{u}$, the restricted residual sum of squares is $RSS_R = \sum(Y - \bar{Y})^2 = \sum y^2$, and we have

$$F = \frac{\sum y^2 - \left(\sum y^2 - \hat{\beta}_1 \sum xy\right)}{\hat{\sigma}^2}$$

$$= \frac{\hat{\beta}_1 \sum xy}{\hat{\sigma}^2} = \frac{\hat{\beta}_1^2 \sum x^2}{\hat{\sigma}^2} = \frac{(\hat{\beta}_1 - 0)^2}{\left(\frac{\hat{\sigma}^2}{\sum x^2}\right)}$$

$$= t_1^2$$

where

$$t_1 = \frac{\hat{\beta}_1}{SE\left(\hat{\beta}_1\right)}$$

is the t-ratio for testing H_0: $\beta_1 = 0$. (This derivation also confirms that $F(1,n-2)=t^2(n-2)$: recall §9.4.)

This approach can also be used to test *joint hypotheses*, for example, H_0: $\beta_0 = 0$, $\beta_1 = 1$: geometrically, the hypothesised line is a 45° line going through the origin. Since the null contains two restrictions, the test statistic (12.5) becomes

$$F = \frac{\left(RSS_R - RSS_U\right)/2}{RSS_U/(N-2)} \sim F(2,N-2)$$

We can use this approach to hypothesis testing on our logarithmic consumption function. The fitted regression line $\ln\hat{C} = 0.087 + 0.982\ln Y$ provides us with $RSS_U = 0.0403$. For the hypothesis H_0: $\beta_1 = 0$, $RSS_R = \sum y^2 = 12.2854$. Thus

$$F = \frac{12.2854 - 0.0403}{0.0403/61} = 18550 > F_{0.05}(1,61) = 4.00$$

so that H_0: $\beta_1 = 0$ is clearly rejected. Note also that the t-ratio here is

$$t_1 = \frac{0.982}{0.007} = 136.2 = \sqrt{F}$$

Let us now consider testing the joint hypothesis H_0: $\beta_0 = 0$, $\beta_1 = 1$ If this was true, then the consumption function would take the form $\ln C = \ln Y + u$. The restricted residual sum of squares is given by $RSS_R = \sum(\ln C - \ln Y) = 1.5812$ and thus

$$F = \frac{(1.5812 - 0.0403)/2}{0.0403/61} = 1167 > F_{0.05}(2,61) = 3.15$$

and there is a clear rejection of H_0: $\beta_0 = 0$, $\beta_1 = 1$: the consumption function is *not* a 45° line going through the origin.

12.6 Prediction from regression

Suppose we have obtained the fitted regression equation

$$\hat{Y} = \hat{\beta}_0 + \hat{\beta}_1 X$$

If we have a given value X_0, the *predicted* value for Y is

$$\hat{Y}_0 = \hat{\beta}_0 + \hat{\beta}_1 X_0$$

The *true* value Y_0 is given by the PRF as

$$Y_0 = \beta_0 + \beta_1 X_0 + u_0$$

so that the *prediction error* is

$$e_0 = \hat{Y}_0 - Y_0 = \left(\hat{\beta}_0 - \beta_0\right) + \left(\hat{\beta}_1 - \beta_1\right) X_0 - u_0$$

It is straightforward to show that predictions are unbiased, $E\left(\hat{Y}_0\right) = Y_0$, by showing that the prediction error has mean zero:

$$E(e_0) = E\left(\hat{Y}_0 - Y_0\right) = E\left(\hat{\beta}_0 - \beta_0\right) + X_0 E\left(\hat{\beta}_1 - \beta_1\right) - E(u_0) = 0$$

The *prediction error variance* can be shown to be[6]

$$Var(e_0) = \sigma^2 \left(1 + \frac{1}{N} + \frac{\left(X_0 - \bar{X}\right)^2}{\sum x^2}\right)$$

This is seen to increase the further X_0 is from $\bar{X}$. A 95% *prediction interval* for Y_0 is thus given by

$$\hat{\beta}_0 + \hat{\beta}_1 X_0 \pm t_{0.025}(N-2)\,\hat{\sigma}\sqrt{1 + \frac{1}{N} + \frac{\left(X_0 - \bar{X}\right)^2}{\sum x^2}}$$

Let us predict salary for the mean value of post-school education, $X_0 = \bar{X} = 5$:

$$\hat{Y}_0 = 15.263 + 2.947 \times 5 = 30 = \bar{Y}$$

which just confirms that the regression goes through the sample means. A 95% prediction interval for Y_0 is

$$30 \pm 2.23 \times 4.56 \times \left(1+\frac{1}{12}\right)^{\frac{1}{2}} = 30 \pm 10.58$$

On the other hand, for $X_0 = 8$

$$Y_0 = 15.263 + (2.947 \times 8) \pm (2.23 \times 4.56)\left(1+\frac{1}{12}+\frac{(8-5)^2}{38}\right)^{\frac{1}{2}}$$

i.e.

$$Y_0 = 38.84 \pm 11.68$$

which has a wider prediction interval. Thus, the prediction interval is at a minimum when $X_0 = \bar{X}$ and increases as X_0 departs from the sample mean *in either direction.*

12.7 Functional forms and non-linearity

Although we have been analysing the *linear* regression model, some non-linear models, as we saw in Chapter 3, can be analysed via transformations. One such model has already been introduced within the context of the consumption function, the *log–linear* specification

$$z = Aw^{\beta_1}$$

or

$$Y = \beta_0 + \beta_1 X$$

where

$$Y = \ln(z) \qquad X = \ln(w) \qquad \beta_0 = \ln(A)$$

It is important to note that the term *linear regression model* refers to models which are linear in the *parameters*, so that, for example, variables appearing as powers, such as X^2, $X^{-1} = 1/X$ and $X^{\frac{1}{2}} = \sqrt{X}$, are all allowed, as well as variables such as sin(X), cos(X) and exp(X) as long as they enter linearly. Popular examples discussed in §3.4 are the *semi-log* models:

$$z = A\exp(\beta_1 X) \qquad \text{and} \qquad \exp(Y) = Aw^{\beta_1}$$

which transform to

$$\ln(Y) = \ln(A) + \beta_1 X \qquad \text{and} \qquad Y = \ln(A) + \beta_1 \ln(w)$$

and the *reciprocal* models

$$Y = \beta_0 + \beta_1\left(\frac{1}{X}\right) \qquad \text{and} \qquad Y = \beta_0 + \beta_1\left(\frac{1}{\sqrt{X}}\right)$$

However, we must be careful when introducing the error term. The model

$$z = Aw^{\beta_1}e^u$$

in which the error enters *multiplicatively*, transforms to

$$\ln(z) = \ln(A) + \beta_1 \ln(w) + u$$

which can be estimated by OLS, whereas the model

$$z = Aw^{\beta_1} + u$$

in which the error enters *additively*, cannot be transformed and must be estimated by a technique known as *non-linear least squares.*

Notes

1 Note the change in notation from slope *a* and intercept *b* to $\hat{\beta}_0$ and $\hat{\beta}_1$ and, below, the residual from e_i to $\hat{u}_i$. This is done both to emphasise that the slope, intercept and residual can, as we shall shortly see, be regarded as estimators of underlying population parameters and to enable a more general notation to be used consistently in subsequent development of the regression model.

2 The results to be presented in subsequent sections of this chapter assume the regressor to be non-stochastic. Independence of the error and regressor is discussed in Chapter 16.

3 The proof that $\hat{\beta}_1$ is BLUE is as follows. Consider any linear estimator $\tilde{\beta}_1 = \sum dy$. For $\tilde{\beta}_1$ to be unbiased we must have

$$E(\tilde{\beta}_1) = \sum dE(y) = \sum dx\beta_1 = \beta_1 \sum dx = \beta_1$$

so that $\sum dx = 1$. Under the classical assumptions, the variance of $\tilde{\beta}_1$ will be $V(\tilde{\beta}_1) = \sum d^2\sigma^2$, and we have to find the *d* that minimises $V(\tilde{\beta}_1)$ subject to $\sum dx = 1$. This is achieved by minimising the expression $\sum d^2 - \lambda(\sum dx - 1)$, where λ is a Lagrangian multiplier. Differentiating with respect to *d* and equating to zero gives $2d - \lambda x = 0$ or $d = (\lambda/2)x$, which implies that

$$\Sigma dx = 1 = \frac{\lambda}{2}\Sigma x^2$$

i.e., that $\lambda = 2/\sum x^2$. Hence $d = x/\sum x^2$ and $\tilde{\beta}_1 = \sum xy/\sum x^2 = \hat{\beta}_1$. Thus the OLS estimator has the minimum variance in the class of linear unbiased estimators, this minimum variance being

$$V(\hat{\beta}_1) = \sum\left(\frac{x}{\sum x^2}\right)^2\sigma^2 = \frac{\sigma^2}{\sum x^2}$$

The alternative estimator in §12.3 sets $d_1 = -1/(x_N - x_1)$ and $d_N = 1/(x_N - x_1)$, with $d_2 = \ldots = d_{N-1} = 0$, where the pairs of observations are sorted in ascending order of *X*. Thus it is clear that $\tilde{\beta}_1$ is unbiased, since

$$\sum dx = -\frac{x_1}{x_N - x_1} + \frac{x_N}{x_N - x_1} = 1$$

Its variance will be given by

$$V(\tilde{\beta}_1) = \frac{2\sigma^2}{(x_N - x_1)^2}$$

so that

$$V(\tilde{\beta}_1) - V(\hat{\beta}_1) = \frac{2\sigma^2}{(x_N - x_1)^2} - \frac{\sigma^2}{x_1^2 + x_2^2 + \ldots + x_N^2}$$
$$= \sigma^2\frac{2(x_2^2 + \ldots + x_N^2) + (x_N - x_1)^2}{(x_N - x_1)^2(x_1^2 + x_2^2 + \ldots + x_N^2)}$$

which is clearly positive. It is, of course, possible that we might find non-linear estimators that are unbiased but have a smaller variance than $\hat{\beta}_1$. However, if the normality assumption 5 holds, then $\hat{\beta}_1$ will have minimum variance amongst *all* unbiased estimators. The OLS estimators are also consistent, since as the sample size increases the denominator of $V(\hat{\beta}_1)$ must also increase.

4 Derivations of these results are as follows. First note that, using the mean deviation form $y = \beta_1 x + u$,

$$\text{Thus } \hat{\beta}_1 = \frac{\sum xy}{\sum x^2} = \frac{\sum x(\beta_1 x + u)}{\sum x^2} = \beta_1 + \frac{\sum xu}{\sum x^2}$$

$$V(\hat{\beta}_0) = E(\hat{\beta}_0 - \beta_0)^2 = E(\bar{Y} - \hat{\beta}_1\bar{X} - \beta_0)^2 = E(\beta_0 + \beta_1\bar{X} + \bar{u} - \hat{\beta}_1\bar{X} - \beta_0)^2$$

$$= E(\bar{u} - (\hat{\beta}_1 - \beta_1)\bar{X})^2 = E\left(\bar{u} - \bar{X}\frac{\sum xu}{\sum x^2}\right)^2$$

$$= \sigma^2\left(\frac{1}{N} + \frac{\bar{X}^2}{\sum x^2}\right)$$

The derivation of $V(\hat{\beta}_0)$ uses the 'population counterpart' of $\bar{Y} = \hat{\beta}_0 + \hat{\beta}_1\bar{X}$, $\bar{Y} = \beta_0 + \beta_1\bar{X} + \bar{u}$, from which, using §10.2, $V(\bar{u}) = E(\bar{u}^2) = \sigma^2/N$ s.

5 This extension also enables us to demonstrate the unbiasedness of $\hat{\sigma}^2$: since $\hat{\sigma}^2 = \sigma^2\chi^2(N-2)/(N-2)$ then $E\hat{\sigma}^2 = \sigma^2$ on using the result from §9.4 that $E(\chi^2(N-2)) = N-2$.

6 The variance of the prediction error can be decomposed as

$$V(e_0) = V(\hat{\beta}_0) + X_0^2 V(\hat{\beta}_1) + 2X_0 Cov(\hat{\beta}_0, \hat{\beta}_1) + V(u_0)$$

$$= \sigma^2\left(\frac{1}{N} + \frac{\bar{X}^2}{\sum x^2}\right) + \sigma^2\frac{X_0^2}{\sum x^2} - 2X_0\sigma^2\frac{\bar{X}}{\sum x^2} + \sigma^2$$

$$= \sigma^2\left(1 + \frac{1}{N} + \frac{(X_0 - \bar{X}^2)}{\sum x^2}\right)$$

13
Multiple Regression

Abstract: *The bivariate regression model of the previous chapter is extended to allow more than one regressor (independent variable) to be included, which leads to certain necessary extensions to statistical inference and hypothesis testing. The possibility that omitting an important variable will lead to biased estimates of the other coefficients is analysed both algebraically and by way of an empirical example, with further examples used to illustrate other aspects of hypothesis testing within a multiple regression framework. The concept of multicollinearity, in which regressors are highly correlated with each other and which has, of course, no counterpart in bivariate regression, is developed by way of a simulation example.*

13.1 Incorporating additional regressors

The 'bivariate' regression model analysed so far is quite restrictive, as it only allows Y to be influenced by a single regressor, X. A moment's reflection about the two examples that we have been using shows how limited this model is: why shouldn't salary be influenced by *both* post-school education and experience? And ought we to consider determinants of consumption other than just income, such as inflation and interest rates?

The question that we address now is how we should go about estimating the parameters of the 'trivariate' model

$$Y_i = \beta_0 + \beta_1 X_{1i} + \beta_2 X_{2i} + u_i, \qquad i = 1,\ldots,N$$

One possibility might be to regress Y on X_1 and X_2 separately, thus obtaining the bivariate regression slope estimators

$$\hat{b}_1 = \frac{\sum x_1 y}{\sum x_1^2} \qquad \hat{b}_2 = \frac{\sum x_2 y}{\sum x_2^2}$$

but *two* intercept estimators

$$\hat{b}_{01} = \bar{Y} - \hat{b}_1 \bar{X}_1 \qquad \hat{b}_{02} = \bar{Y} - \hat{b}_2 \bar{X}_2$$

This *non-uniqueness* of the intercept estimates is just one problem of this approach. An arguably much more serious defect is that, in general, and as we will show later, $\hat{b}_1$ and $\hat{b}_2$ are *biased* estimators of β_1 and β_2, so that $E(\hat{b}_1) \neq \beta_1$ and $E(\hat{b}_2) \neq \beta_2$.

To obtain BLU (best linear unbiased) estimates, we must estimate a *multiple regression* by OLS (ordinary least squares). The PRF (population regression function) now becomes

$$Y_i = \beta_0 + \beta_1 X_{1i} + \beta_2 X_{2i} + u_i$$

and classical assumption 4 of §12.2 generalises to

$$Cov(u_i, x_{1j}) = Cov(u_i, x_{2j}) = E(u_i x_{1j}) = E(u_i x_{2j}) = 0$$

OLS estimates are obtained by minimising

$$\sum \hat{u}_i^2 = \sum (Y_i - \hat{\beta}_0 - \hat{\beta}_1 X_{1i} - \hat{\beta}_2 X_{2i})^2$$

with respect to $\hat{\beta}_0, \hat{\beta}_1$ and $\hat{\beta}_2$. This yields three normal equations, which can be solved to obtain[1]

$$\hat{\beta}_1 = \frac{\sum yx_1 \sum x_2^2 - \sum yx_2 \sum x_1x_2}{\sum x_1^2 \sum x_2^2 - \left(\sum x_1x_2\right)^2}$$

$$\hat{\beta}_2 = \frac{\sum yx_2 \sum x_1^2 - \sum yx_1 \sum x_1x_2}{\sum x_1^2 \sum x_2^2 - \left(\sum x_1x_2\right)^2}$$

$$\hat{\beta}_0 = \bar{Y} - \hat{\beta}_1\bar{X}_1 - \hat{\beta}_2\bar{X}_2$$

It can also be shown that

$$V(\hat{\beta}_1) = \sigma^2 \frac{\sum x_2^2}{\sum x_1^2 \sum x_2^2 - \left(\sum x_1x_2\right)^2}$$

$$V(\hat{\beta}_2) = \sigma^2 \frac{\sum x_1^2}{\sum x_1^2 \sum x_2^2 - \left(\sum x_1x_2\right)^2}$$

$$Cov(\hat{\beta}_1, \hat{\beta}_2) = \sigma^2 \frac{-\sum x_1x_2}{\sum x_1^2 \sum x_2^2 - \left(\sum x_1x_2\right)^2}$$

$$V(\hat{\beta}_0) = \frac{\sigma^2}{N} + \bar{X}_1^2 V(\hat{\beta}_1) + 2\bar{X}_1\bar{X}_2 Cov(\hat{\beta}_1, \hat{\beta}_2) + \bar{X}_2^2 V(\hat{\beta}_2)$$

These formulae all contain σ^2, an estimate of which is now given by

$$\hat{\sigma}^2 = \frac{RSS}{N-3}$$

where now

$$RSS = \sum \hat{u}^2 = \sum y^2 - \hat{\beta}_1 \sum x_1 y - \hat{\beta}_2 \sum x_2 y$$

It then follows that

$$t_i = \frac{\hat{\beta}_i - \beta_i}{SE(\hat{\beta}_i)} \sim t(N-3), \qquad i = 0,1,2$$

Note that the degrees of freedom are N–3 because we now have two regressors. Hypothesis tests and confidence intervals can be constructed in an analogous fashion to those of bivariate regression.

However, with two regressors, further hypotheses are of interest. One in particular is the *joint* hypothesis

$$H_0 : \beta_1 = \beta_2 = 0$$

which is to be tested against the alternative that β_1 and β_2 are *not both zero*. This can be tested using the approach introduced in §**12.5**, being based on the statistic

$$F = \frac{\left(RSS_R - RSS_U\right)/2}{RSS_U / \left(N-3\right)} = \frac{\left(RSS_R - RSS_U\right)/2}{\hat{\sigma}^2} s \sim F\left(2, N-3\right)$$

where

$$RSS_U = RSS = \sum y^2 - \hat{\beta}_1 \sum x_1 y - \hat{\beta}_2 \sum x_2 y$$

and

$$RSS_R = \sum y^2$$

Now, analogous to r_{XY}^2 in bivariate regression, we can define the *coefficient of MULTIPLE determination* as

$$R^2 = \frac{RSS_R - RSS_U}{RSS_R}, \; 0 \le R^2 \le 1$$

and some algebra then yields

$$F = \frac{R^2}{1 - R^2} \cdot \frac{N-3}{2}$$

Salaries, education and experience yet again

Returning to our salaries example, multiple regression yields the following estimates, now denoting education as X_1 and experience as X_2 and noting that the 'common denominator' is

$$\sum x_1^2 \sum x_2^2 - \left(\sum x_1 x_2\right)^2 = 38 \times 30 - 25^2 = 515$$

$$\hat{\beta}_1 = \frac{112 \times 30 - 124 \times 25}{515} = \frac{260}{515} = 0.505$$

$$\hat{\beta}_2 = \frac{124 \times 38 - 112 \times 25}{515} = \frac{1912}{515} = 3.713$$

$$\hat{\beta}_0 = 30 - 0.505 \times 5 - 3.713 \times 10 = -9.650$$

$$\hat{\sigma}^2 = s\frac{538 - 0.505 \times 112 - 3.713 \times 124}{9} = \frac{21.09}{9} = 2.343$$

The variances and standard errors of the slope estimates are thus

$$V(\hat{\beta}_1) = 2.343\frac{30}{515} = 0.136 \qquad SE(\hat{\beta}_1) = 0.369$$

$$V(\hat{\beta}_2) = 2.343\frac{38}{515} = 0.173 \qquad SE(\hat{\beta}_2) = 0.416$$

For the variance and hence standard error of the intercept, we need

$$Cov(\hat{\beta}_1, \hat{\beta}_2) = 2.343\frac{-25}{515} = -0.114$$

so that

$$V(\hat{\beta}_0) = \frac{2.343}{12} + 5^2 \times 0.136 + 2 \times 5 \times 10 \times \left(-0.114\right) + 10^2 \times 0.173 = 9.523$$

and

$$SE(\hat{\beta}_0) = 3.086$$

The t-ratios for testing the individual hypotheses $\beta_1 = 0$, $\beta_2 = 0$ and $\beta_0 = 0$ are

$$t_1 = \frac{0.505}{0.369} = 1.37, \qquad t_2 = \frac{3.713}{0.416} = 8.93$$

and

$$t_0 = -\frac{9.650}{3.086} = -3.13$$

Since $t_{0.025}(9) = 2.26$, both $\beta_2 = 0$ and $\beta_0 = 0$ can be rejected at the 5% level, while $\beta_1 = 0$ cannot be rejected.

Furthermore,

$$R^2 = \frac{538 - 21.09}{538} = 0.9608$$

and

$$F = \frac{0.9608}{1 - 0.9608} \cdot \frac{9}{2} = 110.3$$

As $F_{0.05}(2,9) = 4.26$, we can reject H_0: $\beta_1 = \beta_2 = 0$. The regressors X_1 and X_2 explain 96.1% of the variation in Y.

The estimated multiple regression

$$\underset{}{\hat{Y} = -\underset{(3.086)}{9.650} + \underset{(0.369)}{0.505}\, X_1 + \underset{(0.416)}{3.713}\, X_2}$$

has some interesting features that are worth comparing with the two bivariate regressions:

$$\hat{Y} = \underset{(3.926)}{15.263} + \underset{(0.740)}{2.947}\, X_1$$

and

$$\hat{Y} = -\underset{(2.950)}{11.333} + \underset{(0.291)}{4.133}\, X_2$$

First, the estimates of β_0 and β_2 are close to each other in the multiple and Y on X_2 regressions, but the estimates $\hat{\beta}_1$ and $\hat{b}_1$ of β_1 are very different:

indeed, they are of different orders of magnitude and the former is insignificant. Consequently, it appears that X_1, post-school education, is not a significant factor in determining salaries after all, and it is only experience that counts. Education only appears significant when experience is *excluded*, so that it is acting as a 'proxy' for experience and the Y on X_1 regression is *spurious* (cf. the partial correlation analysis of §5.5).

Second, in the multiple regression we have only nine degrees of freedom, so information is rather limited. This is why we need t-ratios in excess of 2.3 for coefficients to be significant at the 5% level, and this can be quite difficult to achieve. It is often sensible to choose larger significance levels (and hence smaller critical values) for small sample sizes – and, conversely, low significance levels for very large samples (recall the arguments of §11.5). Further, note that the F-statistic rejects H_0: $\beta_1 = \beta_2 = 0$ even though $\beta_1 = 0$ cannot be rejected on a t-test: *just one* of the coefficients needs to be non-zero for H_0 to be rejected.

The spurious nature of the above Y on X_1 regression can be explained algebraically. Comparing the formulae for $\hat{b}_1$ and $\hat{\beta}_1$,

$$\hat{b}_1 = \frac{\sum x_1 y}{\sum x_1^2} \qquad \hat{\beta}_1 = \frac{\sum yx_1 \sum x_2^2 - \sum yx_2 \sum x_1 x_2}{\sum x_1^2 \sum x_2^2 - \left(\sum x_1 x_2\right)^2}$$

we see that $\hat{b}_1 = \hat{\beta}_1$ only if $\Sigma x_1 x_2 = 0$, i.e., only if X_1 and X_2 are *uncorrelated* ($r_{12} = 0$, in which case we will also have $V(\hat{b}_1) = V(\hat{\beta}_1)$). If the two estimates are identical, multiple regression 'collapses' to a pair of simple regressions. In general, though,

$$\hat{b}_1 = \frac{\sum x_1 (\beta_1 x_1 + \beta_2 x_2 + u)}{\sum x_1^2} = \frac{\beta_1 \sum x_1^2 + \beta_2 \sum x_1 x_2 + \sum x_1 u}{\sum x_1^2}$$

$$= \beta_1 + \beta_2 \frac{\sum x_1 x_2}{\sum x_1^2} + \frac{\sum x_1 u}{\sum x_1^2}$$

Thus

$$E\left(\hat{b}_1\right) = \beta_1 + \beta_2 \frac{\sum x_1 x_2}{\sum x_1^2}$$

since the last term in the formula for $\hat{b}_1$ has zero expectation from assumption 4. Hence, if β_2 and $\Sigma x_1 x_2$ have the *same* sign, $E(\hat{b}_1) > \beta_1$ and the *bias* $E(\hat{b}_1)-\beta_1$ is positive; whereas if they are of *opposite* sign, the bias is negative. Two related points are worth noting: (i) $\Sigma x_1 x_2$ is the same sign as the correlation between X_1 and X_2, and (ii) $\Sigma x_1 x_2 / \Sigma x_1^2$ is the slope coefficient in the regression of X_2 on X_1. We can thus explain why we obtained the results that we did from simple regression: X_1 and X_2 are positively correlated ($r_{12} = 0.74$) and, if β_1 is actually zero and β_2 is positive, $E(\hat{b}_1) > 0$, so that obtaining $\hat{b}_1 = 2.947$ when $\beta_1 = 0$ is consistent with theory.

13.2 Regression with *k* explanatory variables

Let us now consider the general multiple regression where we have *k* regressors:

$$Y_i = \beta_0 + \beta_1 X_{1i} + \ldots + \beta_k X_{ki} + u_i$$

OLS estimates $\hat{\beta}_0, \hat{\beta}_1, \ldots, \hat{\beta}_k$ are BLUE under our set of assumptions. We do not provide formulae as they are 'impossible' to obtain algebraically without using a matrix formulation, and hence can only realistically be calculated using an appropriate econometric software package.

Nevertheless, all standard results carry forward, with minor alterations to reflect the number of regressors, for example,

$$t_i = \frac{\hat{\beta}_i - \beta_i}{SE\left(\hat{\beta}_i\right)} \sim t(N-k-1)$$

$$F = \frac{R^2}{1-R^2} \cdot \frac{N-k-1}{k} \sim F(k, N-k-1)$$

$$\hat{\sigma}^2 = \frac{RSS}{N-k-1}$$

We have referred to the quantity $N-k-1$ as the *degrees of freedom* of the regression. Why is this?

Suppose $k = 2$, so that the normal equations are

$$\sum \hat{u} = 0; \quad \sum x_1 \hat{u} = 0; \quad \sum x_2 \hat{u} = 0$$

These three equations *fix* the values of three residuals, so that only $N-3$ are free to vary. Thus, if there are $k=N-1$ regressors, then there are *NO* degrees of freedom and the regression technique breaks down (in practice, it becomes problematic well before this limit is reached).

Including an additional variable in the regression cannot increase the *RSS*, for it will always explain some part of the variation of *Y*, even if by only a tiny (and insignificant) amount, unless the estimated coefficient is *exactly* zero, which is highly unlikely. Hence, from its definition, R^2 will increase towards 1 as more regressors are added, even though they may be both economically and statistically unnecessary. To adjust for this effect, we can define the *R-bar-squared* statistic

$$\bar{R}^2 = 1 - \frac{N-1}{N-k-1}\left(1-R^2\right) \qquad (13.1)$$

$\bar{R}^2$ has some interesting properties. The definition (13.1) can be written as

$$\frac{\left(1-\bar{R}^2\right)}{N-1}\sum y^2 = \frac{\sum y^2}{N-k-1}\left(1-R^2\right)$$

Now recall that $R^2 = (RSS_R - RSS_U)/RSS_R$, $\hat{\sigma}^2 = RSS_U/\left(N-k-1\right)$ and $RSS_R = \sum y^2$. Hence

$$1-R^2 = \frac{N-k-1}{\sum y^2}\hat{\sigma}^2$$

so that

$$\frac{\left(1-\bar{R}^2\right)}{N-1}\sum y^2 = \hat{\sigma}^2$$

Thus the set of regressors that minimises $\hat{\sigma}^2$ will be the set that maximises $\bar{R}^2$. Note also that if $R^2 < k/(N-1)$ then $1-R^2 > (N-k-1)/(N-1)$ and, from (13.1), $\bar{R}^2 < 0$. For example, with $k = 4$ regressors and $N = 21$ observations then, if $R^2 < 0.2$, $\bar{R}^2$ will be negative.

13.3 Hypothesis tests in multiple regression

There are a variety of hypotheses in multiple regression models that we might wish to consider. All can be treated within the general framework of the F-test by interpreting the null hypothesis as a set of (linear) restrictions imposed on the regression model which we wish to test to find out whether they are *acceptable*:

$$F = \frac{RSS_R - RSS_U}{RSS_U} \cdot \frac{N-k-1}{r} = \frac{\left(RSS_R - RSS_U\right)/r}{\hat{\sigma}^2} \sim F\left(r, N-k-1\right)$$

where

- RSS_U: RSS from the *unrestricted* regression, i.e., the regression estimated under the alternative hypothesis (without restrictions).
- RSS_R: RSS from the *restricted* regression, i.e., the regression estimated under the null hypothesis (with restrictions imposed).
- r: the number of restrictions imposed *by the null* hypothesis.

An equivalent form of the test statistic, which may sometimes be easier to compute, is

$$F = \frac{R_U^2 - R_R^2}{1 - R_U^2} \cdot \frac{N-k-1}{r}$$

where R_U^2 and R_R^2 are the R^2s from the unrestricted and restricted regressions respectively.

Some examples of hypotheses that might be investigated are the following:

1. We might be interested in testing whether a *subset* of the coefficients are zero, for example,

$$H_0 : \beta_{k-r+1} = \beta_{k-r+2} = \ldots = \beta_k = 0$$

that is, that the last r regressors are *irrelevant* (the theory that suggests including them in the model is false). Here the *restricted* regression is one that contains the first $k-r$ regressors (note that the ordering of the regressors is at our discretion):

$$Y_i = \beta_0 + \beta_1 X_{1i} + \ldots + \beta_{k-r} X_{k-r,i} + u_i$$

The test statistic here can be written as

$$rF = \frac{\left(N-k+r-1\right)\hat{\sigma}_R^2 - \left(N-k-1\right)\hat{\sigma}_U^2}{\hat{\sigma}_U^2}$$

where $\hat{\sigma}_U^2 = RSS_U/(N-k-1)$ and $\hat{\sigma}_R^2 = RSS_R/(N-k+r-1)$. This statistic can be solved as

$$\frac{\hat{\sigma}_R^2}{\hat{\sigma}_U^2} = \frac{\alpha+F}{\alpha+1} \quad \text{where} \quad \alpha = \frac{N-k-1}{r}$$

Thus $\hat{\sigma}_R^2 \gtreqless \hat{\sigma}_U^2$ according to whether $F \gtreqless 1$, which implies that if the F-statistic associated with a set of regressors is less than 1 then dropping these regressors will reduce the error variance and hence increase $\bar{R}^2$. For the single restriction case of $r=1$, this means that if the absolute value of the t-ratio for a regressor is less then 1 then dropping this regressor will reduce $\hat{\sigma}^2$ and increase $\bar{R}^2$.

2. A more complicated type of restriction is where the coefficients obey a *linear restriction* of the general form

$$H_0:\ c_1\beta_1 + c_2\beta_2 + \ldots + c_k\beta_k = d$$

where $c_1, c_2, \ldots, c_k, d$ are constants. An example of this type of restriction, which occurs regularly in economics, is two coefficients summing to zero, for example, H_0: $\beta_1 + \beta_2 = 0$. This is obtained from the general form by setting $c_1 = c_2 = 1, c_3 = \ldots .c_k = 0$, $d = 0$. To construct a test of this hypothesis we have to be able to estimate the restricted regression. Suppose $k=2$ for simplicity. The restricted model is then, since the hypothesis implies $\beta_2 = -\beta_1$,

$$Y = \beta_0 + \beta_1 X_1 - \beta_1 X_2 + u$$

or

$$Y = \beta_0 + \beta_1 (X_1 - X_2) + u$$

so that the restricted model is the regression of Y on $X^* = X_1 - X_2$. An equivalent test is the t-ratio on X_2 in the regression of Y on X^* and X_2: suppose this regression is written

$$\begin{aligned} Y &= \gamma_0 + \gamma_1 X^* + \gamma_2 X_2 + u \\ &= \gamma_0 + \gamma_1 X_1 + (\gamma_2 - \gamma_1) X_2 + u \end{aligned}$$

Thus only if $\gamma_2 = 0$ will the coefficients on X_1 and X_2 sum to zero, and this can be tested by the t-ratio on X_2 in the regression of Y on X^* and X_2.

The relationships between t-ratios for individual regressors and F-statistics for sets of regressors are often subtle and need to be treated carefully. It is possible that although the F-statistic exceeds 1, all the t-ratios are less than 1, which may occur when the regressors are highly inter-correlated (this is known as *multicollinearity* and will be discussed in §13.4). This does not mean that $\bar{R}^2$ can be increased by dropping *all* the regressors, since the dropping of one regressor will alter the other t-ratios.

It is also possible for the F-statistic to be less than one yet for all of the t-ratios to exceed one. This is a rather complicated situation for which some guidelines are available. There is also an interesting result that states that if a regressor is omitted then there can be no change in the sign of the coefficient on any regressor that is more significant (has a higher absolute t-ratio) than the omitted regressor.[2]

Modelling an extended consumption function

We now return to our consumption function example and consider including two further regressors, inflation and the (long) interest rate. To help with notation, we now denote the logarithms of consumption and income by c_t and y_t respectively, inflation by π_t and the interest rate by R_t. The following extended consumption regression was fitted:

$$\underset{}{\hat{c}_t} = \underset{(0.0005)}{0.9917\ y_t} + \underset{(0.00072)}{0.00212\ \pi_{t-1}} - \underset{(0.00199)}{0.00453\ R_t} - \underset{(0.00204)}{0.00336\ R_{t-1}}$$

$$R^2 = 0.9985 \qquad RSS = 0.017362 \qquad \hat{\sigma} = 0.01730$$

Recall from §12.4 that the intercept was insignificant in the bivariate consumption function: it remains so in the multiple regression and hence is excluded. Lagged inflation π_{t-1} and both R_t and R_{t-1} are included as regressors: consumption is thus seen to depend upon *last year's* rate of inflation and both *current and last year's* levels of the interest rate. Inflation positively influences consumption because higher inflation reduces purchasing power and so induces an increase in current consumption at the expense of deferred consumption. Increases in interest rates, on the other hand, encourages saving at the expense of current consumption; hence the negative coefficients on the interest rate regressors. The presence of lagged regressors emphasises the dynamic nature of the typical empirical consumption function.

The coefficient on y_t, although very close to 1, is significantly below this value: a test of H_0: $\beta_1 = 1$ against H_A: $\beta_1 < 1$ is provided by the test statistic $t = (0.99175-1)/0.00046 = -17.93 \sim t(58)$, which is highly significant. Both π_{t-1} and R_t are clearly significant, but there is some doubt over R_{t-1}, whose t-ratio is just −1.645 with a p-value of 0.105. However, the coefficients on R_t and R_{t-1} are reasonably similar, suggesting the linear restriction $\beta_3 - \beta_4 = 0$.

A test of this hypothesis is provided by estimating the restricted regression

$$c_t = \beta_1 y_t + \beta_2 \pi_{t-1} + \beta_3 (R_t + R_{t-1}) + u_t$$

This yields $RSS_R = 0.017389$ and hence the test statistic

$$F = \frac{0.017389 - 0.017362}{0.01730^2} = 0.09 \sim F(1,58)$$

which is clearly insignificant. The estimated restricted regression is

$$\underset{}{\hat{c}_t} = \underset{(0.0005)}{0.9917}\ y_t + \underset{(0.00071)}{0.00214}\ \pi_{t-1} - \underset{(0.00109)}{0.00791}\ (R_t + R_{t-1})/2$$

$$R^2 = 0.9985 \qquad \hat{\sigma} = 0.01717$$

Since the F-statistic associated with the restriction is less than 1, the error variance has declined and β_3 now enters significantly (note that we have interpreted the restriction in terms of the *average* interest rate over two years, so that the reported coefficient is $2\hat{\beta}_3$).

There is an interesting link between t-ratios and the concept of partial correlations introduced in §5.5. Suppose we have the set of variables $Y, X_1, \ldots, X_k$. The partial correlation between Y and X_1 with $X_2, \ldots, X_k$ held constant may be denoted $r_{Y1.2\cdots k}$. Then, if t_1 is the t-ratio on X_1 in the regression of Y on $X_1, \ldots X_k$, it can be shown that[3]

$$r^2_{Y1\cdot2\cdots k} = \frac{t_1^2}{t_1^2 + (N-k-1)} \qquad \text{or} \qquad t_1^2 = (N-k-1)\frac{r^2_{Y1\cdot2\cdots k}}{1 - r^2_{Y1\cdot2\cdots k}}$$

Thus the partial correlations between consumption and income, inflation and average interest rates, respectively, may be calculated as

$$r_{cy\cdot\pi R} = \sqrt{\frac{2197.547^2}{2197.547^2 + 59}} = 0.9999$$

$$r_{c\pi\cdot yR} = \sqrt{\frac{3.01645^2}{3.01645^2 + 59}} = 0.366$$

$$r_{cR\cdot y\pi} = -\sqrt{\frac{7.24777^2}{7.24777^2 + 59}} = -0.686$$

Thus, although there is an extremely high partial correlation between consumption and income, inflation and interest rates nevertheless still enter as important determinants of consumption.

13.4 Multicollinearity

When estimating multiple regressions, there is an important problem that is often encountered that can have no counterpart in bivariate regression, that of *multicollinearity.* We investigate this problem by employing another simulation example. Suppose that the 'true' PRF is

$$Y_i = 10 + 5X_{1i} + 2X_{2i} + u_i \qquad u_i \sim IN\left(0,100\right)$$

and that the two regressors are correlated. To 'design' this correlation , the following *generating process* for X_2 is used:

$$X_{2i} = 5 + 2X_{1i} + \varepsilon_i \qquad \varepsilon_i \sim IN\left(0,\sigma_\varepsilon^2\right)$$

with $X_{1i} = i$, $i = 1,\ldots,100$. The choice of error variance, σ_ε^2, is at our disposal. We now investigate what happens when we reduce σ_ε^2: that is, what happens when we make the relationship between X_1 and X_2 'tighter and tighter'.

The regression estimates obtained as σ_ε^2 is reduced are shown in Table 13.1. For $\sigma_\varepsilon^2 \geq 1$ the coefficient estimates are reasonably well estimated and close to their 'true' values of 10, 5 and 2, respectively. For $\sigma_\varepsilon^2 = 0.5^2 = 0.25$, so that $r_{12} \approx 1$, all the estimates become individually insignificant, yet the R^2 is unaffected and implies a strong relationship between Y and the regressors: X_1 and X_2 are therefore *jointly*, but not

TABLE 13.1 *Regression estimates as σ_ε^2 is reduced*

σ_ε	r_{12}	$\hat{\beta}_0$	$\hat{\beta}_1$	$\hat{\beta}_2$	R^2
4	0.99733	10.41 (2.30)	5.44 (0.45)	1.78 (0.23)	0.9987
3	0.99883	5.80 (3.16)	4.05 (0.79)	2.49 (0.40)	0.9983
2	0.99959	13.63 (3.95)	5.73 (1.26)	1.62 (0.63)	0.9985
1	0.99990	11.94 (5.70)	5.01 (2.18)	1.98 (1.09)	0.9988
0.5	0.99996	1.80 (10.13)	2.37 (3.97)	3.33 (1.99)	0.9985

individually, significant (recall the discussion in the previous section). What is going on here?

We can analyse the problem analytically using the standard least squares formulae

$$\hat{\beta}_1 = \frac{\sum yx_1 \sum x_2^2 - \sum yx_2 \sum x_1x_2}{\sum x_1^2 \sum x_2^2 - \left(\sum x_1x_2\right)^2}$$

and

$$V\left(\hat{\beta}_1\right) = \sigma^2 \frac{\sum x_2^2}{\sum x_1^2 \sum x_2^2 - \left(\sum x_1x_2\right)^2}$$

Both have the same denominator, which can be written as

$$\sum x_1^2 \sum x_2^2 - \left(\sum x_1x_2\right)^2 = \left(\frac{\sum x_1^2 \sum x_2^2}{\left(\sum x_1x_2\right)^2} - 1\right)\left(\sum x_1x_2\right)^2$$

$$= \left(\frac{1}{r_{12}^2} - 1\right)\left(\sum x_1x_2\right)^2 \to 0 \quad \text{as} \quad r_{12} \to 1$$

Thus, as the correlation between the regressors increases, so the denominator in the expressions for both the coefficient estimate and its standard error becomes very small. Minor changes in value will thus get magnified into large effects, and hence both estimates and standard errors can become very unstable. In general, if we have the model

$$Y = \beta_0 + \beta_1 X_1 + \beta_2 X_2 + u$$
$$X_2 = \gamma_0 + \gamma_1 X_1 + \varepsilon$$

where $u \sim IN\left(0, \sigma^2\right)$ and $\varepsilon \sim IN\left(0, \sigma_\varepsilon^2\right)$, then we can write

$$\begin{aligned} Y &= \beta_0 + \beta_1 X_1 + \beta_2 (\gamma_0 + \gamma_1 X_1 + \varepsilon) + u \\ &= (\beta_0 + \beta_2 \gamma_0) + (\beta_1 + \beta_2 \gamma_1) X_1 + (u + \beta_2 \varepsilon) \end{aligned}$$

As $\sigma_\varepsilon^2 \to 0$, so that $r_{12} \to 1$, the model collapses to a regression of Y on X_1 alone, and X_2 becomes irrelevant (this can be seen by noting that in the limit ε has both a zero mean and a zero variance, and hence is *always* zero).

Having demonstrated how multicollinearity can arise in a simulated example, the obvious question is: how prevalent is it in actual economic data? The answer is that it all depends on the type of data you are using. With cross-sectional data the problem may not be too important, but with time series data that contain *trends* it may become a serious problem. However, its effects can often be mitigated by the judicious use of transformations, or by dropping regressors that are not central to the aims of the analysis. Wherever possible, greater information, typically more data, will help matters.

Notes

1 Nothing essential is gained by providing the derivation of these equations, which follows an obvious extension of the bivariate regression derivation in §12.1. For those interested, details may be found in G.S. Maddala, *Introduction to Econometrics*, 3rd edition (Wiley, 2001), pp. 129–132. It is here that Maddala introduces the salary, education and work experience example but uses only the first five observations of the twelve used here. I have taken the opportunity of artificially increasing the number of observations by seven to provide a slightly richer sequence of examples.

2 See Maddala, *ibid.*, pp. 164–168, for such guidelines and for an application of this result.

3 See Maddala, *ibid.*, pp. 146–147.

14
Autocorrelation

Abstract: *Basic econometrics essentially deals with violations to the classical assumptions. These occur regularly when analysing economic data, and the first violation we consider is that of dependence between errors in a time series regression. This is known as autocorrelation and leads to inefficient, and in some cases biased, estimates of the regression coefficients. It is thus very important to be able to test for the presence of autocorrelation, for which the standard statistic is that of Durbin and Watson. Estimation with autocorrelated errors is discussed using a detailed example concerning the UK consumption function, and further extensions for when a lagged dependent variable is included as a regressor are considered. The possibility of autocorrelation being a consequence of a misspecified model is also investigated.*

14.1 Econometrics and the invalidity of the classical assumptions

If the assumptions of classical linear regression set out in §12.2 were always satisfied, then much economic data analysis would simply be an exercise in regression. Unfortunately, in many applications of regression to economic data, at least one of the assumptions is found to be invalid, which often makes OLS (ordinary least squares) a sub-optimal estimator. The subject of *econometrics* has thus developed a set of techniques that are designed to test the validity of each of the regression assumptions and to provide extensions when they are found to be invalid.[1] We will therefore look at each of the classical assumptions in turn, beginning with probably the most important for economic applications, assumption 3, that of uncorrelated errors.[2]

14.2 Autocorrelated errors

Assumption 3 of classical linear regression states that

$$Cov\left(u_i, u_j\right) = E\left(u_i u_j\right) = 0 \quad \text{for all } i \neq j$$

i.e., that any two errors have zero covariance and hence are uncorrelated. This assumption may often be false when dealing with time series data. Let us move explicitly to a time series framework by using t subscripts, and consider the following model

$$Y_t = \beta_0 + \beta_1 X_t + u_t \qquad t = 1, 2, \ldots, T \tag{14.1}$$

If the errors are *autocorrelated* (of the first order), then they are generated by the *first-order autoregression*

$$u_t = \rho u_{t-1} + \varepsilon_t \qquad -1 < \rho < 1 \tag{14.2}$$

where $\varepsilon_t \sim IN(0, \sigma_\varepsilon^2)$ (a more detailed treatment of autoregressions is given in Chapter 18). The restriction that ρ is less than one in absolute value is a technical condition that need not concern us for the moment. It can be shown that, for this process, the covariance between the error at time t, u_t, and the error j time periods in the past, u_{t-j}, is[3]

$$E(u_t u_{t-j}) = \rho^j \sigma^2 = \frac{\rho^j \sigma_\varepsilon^2}{1-\rho^2}$$

which is obviously non-zero as long as ρ is non-zero.

14.3 Consequences of autocorrelation

Suppose we estimate (14.1) by OLS and simply ignore (14.2). We will concentrate on the properties of the slope estimator: a similar analysis holds for the intercept. Because assumption 4 still holds, $\hat{\beta}$ remains unbiased but will now have a variance that differs from the OLS variance. If we denote the variance under autocorrelated errors as $V(\hat{\beta}_1)_\rho$, then it can be shown that[4]

$$V(\hat{\beta}_1)_\rho = \frac{\sigma^2}{\sum x^2} \cdot \kappa = V(\hat{\beta}_1) \cdot \kappa$$

where

$$\kappa = 1 + 2\rho \frac{\sum_{t=1}^{T-1} x_t x_{t+1}}{\sum_{t=1}^{T} x_t^2} + 2\rho^2 \frac{\sum_{t=1}^{T-2} x_t x_{t+2}}{\sum_{t=1}^{T} x_t^2} + \ldots + 2\rho^{T-1} \frac{x_1 x_T}{\sum_{t=1}^{T} x_t^2}$$

The 'correction factor' κ can be written approximately as

$$\kappa = 1 + 2\rho\hat{\gamma}_1 + 2\rho^2\hat{\gamma}_2 + \cdots + 2\rho^{T-1}\hat{\gamma}_{T-1}$$

where the $\hat{\gamma}_j$, $j = 1, \ldots, T-1$ are OLS slope estimates from the set of regressions $x_{t+j} = \gamma_j x_t + v_{jt}$. Thus, if $\rho = 0$ then $\kappa = 1$ and no correction is needed because there is no autocorrelation.

However, a typical case with economic data is that both the errors and the regressor will be positively autocorrelated, so that $\rho > 0$ and $\hat{\gamma}_j > 0$. Hence $\kappa > 1$ and $V(\hat{\beta}_1)_\rho$ will be bigger than $V(\hat{\beta}_1)$, often by an appreciable amount. For example, suppose that x_t also follows a first-order autoregression, $x_t = \gamma x_{t-1} + v_t$, so that $E(x_t x_{t-j}) = \gamma^j \sigma_x^2$, where σ_x^2 is the variance of x_t. It is then easy to see that $\hat{\gamma}_j = \hat{\gamma}^j$ and

$$\begin{aligned}
\kappa &= 1 + 2\rho\hat{\gamma} + 2\rho^2\hat{\gamma}^2 + \ldots + 2\rho^{T-2}\hat{\gamma}^{T-2} \\
&= 1 + 2\rho\hat{\gamma}\left(1 + \rho\hat{\gamma} + \rho^2\hat{\gamma}^2 + \ldots + \rho^{T-1}\hat{\gamma}^{T-1}\right) \\
&= 1 + 2\rho\hat{\gamma}\frac{\left(1 - \rho^{T-1}\hat{\gamma}^{T-1}\right)}{1 - \rho\hat{\gamma}}
\end{aligned}$$

Since $-1<\rho,\ \hat{\gamma}<1$ then, for T large, $\rho^{T-1}\hat{\gamma}^{T-1}$ can be regarded as approximately zero so that

$$\kappa = 1+\frac{2\rho\hat{\gamma}}{1-\rho\hat{\gamma}} = \frac{1+\rho\hat{\gamma}}{1-\rho\hat{\gamma}}$$

Thus, if $\hat{\gamma}=\rho=0.8$, then $\kappa = 1.64/0.36 = 4.56$, and ignoring this factor will lead to $V(\hat{\beta}_1)$ underestimating $V(\hat{\beta}_1)_\rho$ by 78%. Actually, in practice the underestimation will be even greater. As we have seen, σ^2 is typically estimated by $\hat{\sigma}^2$, which is only unbiased if $\rho = 0$. If there is autocorrelation, so that $\rho \neq 0$, it can be shown that $\hat{\sigma}^2$ becomes biased, since[5]

$$E\left(\hat{\sigma}^2\right) = \left(T - \frac{1+\rho\gamma}{1-\rho\gamma}\right)\frac{\sigma^2}{T-2}$$

Thus if $T=20$, $E(\hat{\sigma}^2) = 0.86\sigma^2$ and there is a further underestimation of 14%. The overall effect is an underestimation of the standard error of over 80%.

The problem here is that, unless we know that the errors are autocorrelated and have available a value for ρ, we shall continue to use the incorrect OLS variance. As one might expect from the above example, this can cause great problems when carrying out significance tests, as the following simulation example illustrates. Suppose we generate Y and X by

$$X_t = 0.95X_{t-1} + v_t,\quad v_t \sim IN\,(0,1),\quad X_0 = 0$$
$$u_t = 0.95u_{t-1} + \varepsilon_t,\quad \varepsilon_t \sim IN\,(0,1),\quad u_0 = 0$$

and

$$Y_t = u_t$$

that is, we have $\beta_0 = \beta_1 = 0$ in the true relationship, since Y_t *does not* depend on X_t. $T=50$ observations on Y and X are simulated, these being shown in Figure 14.1.

We now estimate the OLS regression of Y_t on X_t, obtaining

$$\begin{aligned} \hat{Y}_t = \ & \underset{(0.192)\ [6.27]}{1.205} + \underset{(0.104)\ [5.43]}{0.565}\, X_t \qquad R^2 = 0.38 \end{aligned} \tag{14.3}$$

We should, of course, have obtained intercept and slope estimates which were insignificant, and an R^2 that was close to zero – yet we get

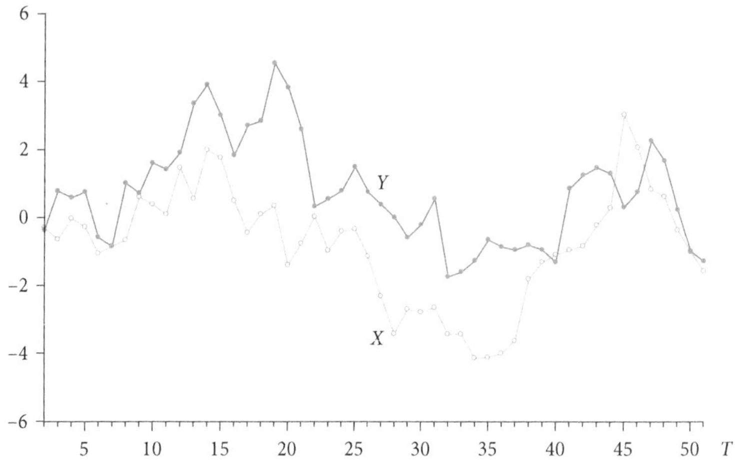

FIGURE 14.1 *50 simulated observations on Y and X*

significantly positive $\hat{\beta}_0$ and $\hat{\beta}_1$ and an R^2 equal to 0.38! What we have got here is another example of a *spurious regression*, where we think that we have uncovered an important relationship when in reality it simply *does not exist*. The reason for the observed correlation between Y_t and X_t, which is 0.62, is that the processes generating the two series introduce a smoothness into them, and this is a very common occurrence in economics (recall the consumption–income example in §5.5).[6]

14.4 Testing for autocorrelation

Given the strong possibility of spurious regressions between economic time series, we must be able to detect such a problem should it exist. As with most regression 'misspecifications', clues can be obtained by examining the residuals. A plot of the residuals from the spurious regression (14.3), shown as Figure 14.2, reveals a distinct *pattern*, there being *runs* of positive and negative residuals.

Such a pattern implies that the residuals cannot be uncorrelated, must therefore have non-zero covariances, and hence exhibit autocorrelation. This pattern is in fact indicative of *positive* autocorrelation, which should be expected from the design of the experiment (ρ was set equal to 0.95).

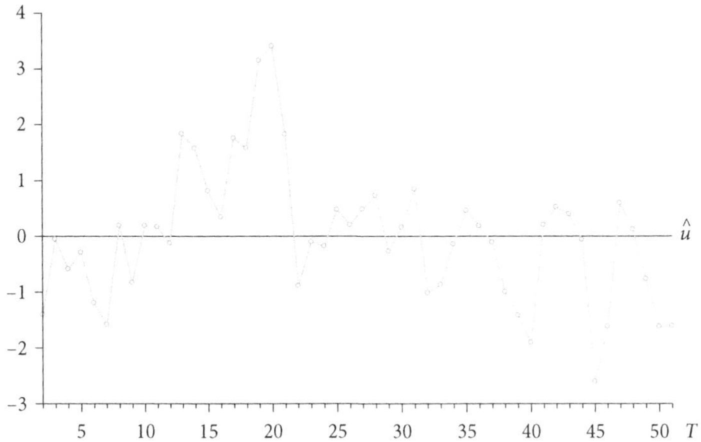

FIGURE 14.2 *Residuals from the spurious regression (14.3)*

Examining a residual plot, while often extremely useful, does not provide a formal *test* for the presence of autocorrelation, that is, a test of H_0: $\rho=0$ in (14.2). The most common test for residual autocorrelation is that of *Durbin and Watson*.[7] The Durbin–Watson statistic, denoted *dw*, assumes that the errors are generated by (14.2), and uses the residuals $\hat{u}_t$ from OLS estimation of (14.1):

$$dw = \frac{\sum_{t=2}^{T}(\hat{u}_t - \hat{u}_{t-1})^2}{\sum_{t=1}^{T}\hat{u}_t^2} = \frac{\sum \hat{u}_t^2 + \sum \hat{u}_{t-1}^2 - 2\sum \hat{u}_t\hat{u}_{t-1}}{\sum \hat{u}_t^2}$$

Since $\sum \hat{u}_t^2$ and $\sum \hat{u}_{t-1}^2$ are approximately equal if the sample size T is large,

$$dw \approx 2\left(1 - \frac{\sum \hat{u}_t\hat{u}_{t-1}}{\sum \hat{u}_{t-1}^2}\right) = 2(1-\hat{\rho})$$

since, by analogy to OLS regression, $\sum \hat{u}_t\hat{u}_{t-1} / \sum \hat{u}_{t-1}^2$ can be regarded as an estimate of ρ. Now, if $\hat{\rho}=0$, $dw=2$, while if $\hat{\rho}$ takes its maximum possible value of + 1, $dw=0$, and if it takes its minimum possible value of −1, $dw=4$. Thus $0 \le dw \le 4$, and the null hypothesis of zero autocorrelation, H_0: $\rho=0$, corresponds to $dw=2$, with the alternative H_A: $\rho>0$ implying that $dw<2$.

How, though, can we obtain critical values to perform hypothesis tests, that is, how far below 2 does *dw* need to be before we can reject the null in favour of the alternative H_A: $\rho > 0$? Unfortunately, the sampling distribution of dw depends upon the number k of regressors, the values that the regressors take, the sample size, and whether a constant is included in the regression or not! This means that rather than having a single critical value corresponding to, say, the 5% level, we actually have a set of *upper* and *lower bounds*, which we denote d_U *and* d_L. Our testing procedure thus becomes

If $dw < d_L$, *REJECT* H_0: $\rho = 0$
If $dw > d_U$, do not reject
If $d_L \leq dw \leq d_U$, the test is *inconclusive*

When T is small and k is large, the 'inconclusive region' can be wide. However, a very useful rule of thumb with most economic data is to treat d_U as *the* critical value. For our spurious regression example, $dw = 0.7$. With $T = 50$, $k = 1$ and a constant included, $d_L = 1.50$ and $d_U = 1.59$ at the 5% level, and thus there is a clear rejection of the null of zero autocorrelation, as we might have expected.

The alternative H_A': $\rho < 0$ implies that $2 < dw < 4$. To test for this alternative of negative autocorrelation, we may simply replace dw by $4-dw$ in the testing procedure above.

14.5 Estimation with autocorrelated errors

Suppose we again have the model (14.1) and (14.2), now written in deviations-from-mean form for simplicity,

$$y_t = \beta x_t + u_t \tag{14.4}$$

$$u_t = \rho u_{t-1} + \varepsilon_t \tag{14.5}$$

We have already shown that OLS is inappropriate. The correct estimation procedure is called generalised *least squares* (GLS). Without going into detail, the BLUE for β is

$$\hat{\beta}_{GLS} = \frac{\sum_{t=2}^{T}(x_t - \rho x_{t-1})(y_t - \rho y_{t-1})}{\sum_{t=2}^{T}(x_t - \rho x_{t-1})^2} + C$$

with variance

$$Var\left(\hat{\beta}_{GLS}\right)=\frac{\sigma^2}{\sum_{t=2}^{T}\left(x_t-\rho x_{t-1}\right)^2}$$

where C and D are correction factors that can be ignored in practice for all but the smallest sample sizes. Note that if $\rho=0$ these formulae collapse to those of OLS. Of course, ρ is typically unknown and must be estimated along with β, and there are various methods of doing this. The general procedure is to note that by substituting (14.4) into (14.5) we have

$$\left(y_t-\beta x_t\right)=\rho\left(y_{t-1}-\beta x_{t-1}\right)+\varepsilon_t$$

or, on rearranging,

$$\left(y_t-\rho y_{t-1}\right)=\beta\left(x_t-\rho x_{t-1}\right)+\varepsilon_t$$

Since the error ε_t obeys all the assumptions of classical linear regression, we can regress the *quasi-differences*, that is, regress $y_t-\rho y_{t-1}$ on $x_t-\rho x_{t-1}$ (the GLS formulae above are of this form). But to do this requires an estimate of ρ, and this can be obtained as (recalling the definition of the *dw* statistic)

$$\hat{\rho}=\frac{\sum\hat{u}_t\hat{u}_{t-1}}{\sum\hat{u}_{t-1}^2}$$

from the regression of $\hat{u}_t$ on $\hat{u}_{t-1}$. Such procedures are fully automatic in econometric packages, a popular version being known as the *Cochrane–Orcutt* technique.[8]

One interesting case occurs when $\rho=1$, because then we have an equation in *first differences*,

$$\Delta y_t=\beta\Delta x_t+\varepsilon_t$$

Jointly estimating our spurious regression system obtains

$$\begin{aligned}&Y_t=\underset{\substack{(0.593)\\[1.42]}}{0.844}+\underset{\substack{(0.159)\\[0.93]}}{0.148}\;X_t+\hat{u}_t, \qquad \hat{u}_t=\underset{(0.099)}{0.768}\;\hat{u}_{t-1}+\varepsilon_t\end{aligned}$$

Since $\hat{\rho}=0.768$ is reasonably close to unity, we might also estimate the first difference regression

$$\Delta \hat{Y}_t = \underset{\substack{(0.158) \\ [0.17]}}{0.026}\ \Delta X_t \qquad R^2 = 0.0002$$

Both regressions confirm the spuriousness of the relationship between Y and X: note the insignificance of the coefficient estimates and the almost zero value of R^2 in the first-difference regression.

Autocorrelation in the consumption function

It is important to emphasise that although spurious regression problems can arise in econometrics, autocorrelation *does not necessarily* imply a spurious relationship. As an example of a 'non-spurious' regression exhibiting autocorrelation, let us return to our consumption function example from §13.3, in which we obtained the regression model

$$\hat{c}_t = \underset{(0.0005)}{0.9917}\,y_t + \underset{(0.00071)}{0.00214}\,\pi_{t-1} - \underset{(0.00109)}{0.00791}\left(R_t + R_{t-1}\right)/2$$

$$R^2 = 0.9985 \qquad \hat{\sigma} = 0.01717$$

For this regression, $dw = 0.54$. With $T = 62$ and $k = 3$, the 5% lower and upper bounds are $d_L = 1.49$, $d_U = 1.69$, so that we have evidence of (positive) residual autocorrelation. Estimating with an autoregressive error yields

$$\hat{c}_t = \underset{(0.0014)}{0.9884}\,y_t + \underset{(0.00055)}{0.00163}\,\pi_{t-1} - \underset{(0.00209)}{0.00184}\left(R_t + R_{t-1}\right)/2$$

$$R^2 = 0.9994 \qquad \hat{\sigma} = 0.0119 \qquad \hat{\rho} = 0.871\,(0.066)$$

Note that both R^2 and $\hat{\sigma}$ improve, indicating a much better overall fit – *but* the standard errors are *larger* on the income and interest rate coefficients. This is not a drawback, for it simply shows that the OLS standard errors were too small, and hence t-ratios too large, so that the estimates were *erroneously precise*: indeed, the interest rate variable is no longer significantly different from zero.

14.6 Testing for autocorrelation with a lagged dependent variable

We very often encounter models in which the *lagged dependent variable*, Y_{t-1}, appears as a regressor:

$$Y_t = \beta_0 + \beta_1 X_t + \alpha Y_{t-1} + u_t$$

As we shall see in **§16.4**, the inclusion of Y_{t-1}, while having no adverse effects on the properties of OLS estimators when the error u_t is independent, may nevertheless have severe consequences if u_t is autocorrelated, for then OLS produces biased estimates of the coefficients.

It is therefore important to test for residual autocorrelation in such models; but an added difficulty is that in these circumstances *dw* is biased towards 2.[9] This means that we could incorrectly accept the null of no autocorrelation simply because *dw* is larger than it should be.

In these circumstances, Durbin's *h*-test should be used:[10]

$$h = \left(1 - \tfrac{1}{2}dw\right)\sqrt{\frac{T}{1 - TV\left(\hat{\alpha}\right)}} \sim N\left(0, 1\right)$$

As an example, consider the following dynamic specification of the consumption function

$$\begin{aligned} \hat{c}_t = {} & 0.6497\,y_t - 0.4675\,y_{t-1} - 0.00063\,\pi_t + 0.8165\,c_{t-1} \\ & (0.0679) \quad (0.0911) \quad (0.00032) \quad (0.0569) \end{aligned}$$

from which $dw = 1.46$. This is just above $d_L = 1.45$, the 5% lower bound for $T = 62$ and $k = 4$, so we might possibly be tempted to conclude that there is no residual autocorrelation. But, on calculating the *h*-test, we obtain

$$h = \left(1 - \tfrac{1}{2} \times 1.46\right)\sqrt{\frac{62}{1 - 62 \times 0.0569^2}} = 2.37$$

Since *h* is distributed as a standard normal, this offers a clear indication of residual autocorrelation (the *p*-value is 0.024), thus rendering both the coefficient estimates and their standard errors unreliable.

14.7 Residual autocorrelation as a consequence of model mis-specification

Why might autocorrelation appear in the residuals? It could, of course, be that the true error process is such that the error at time t is related to the previous error at t–1. An often more useful, and more plausible, explanation is that the appearance of autocorrelation in the residuals is a manifestation of a misspecified model: a variable that has a systematic influence on Y has been omitted, and this influence has been transferred to the residual.

Consider the system

$$y_t = \beta x_t + u_t$$

$$u_t = \rho u_{t-1} + \varepsilon_t$$

or

$$y_t - \beta x_t = \rho\left(y_{t-1} - \beta x_{t-1}\right) + \varepsilon_t$$

This can be rewritten as

$$y_t = \beta x_t - \beta\rho x_{t-1} + \rho y_{t-1} + \varepsilon_t$$

Note that the coefficient on x_{t-1} is (minus) the product of the coefficients on x_t and y_{t-1}. Thus the model is the same as

$$y_t = \gamma_1 x_t + \gamma_2 x_{t-1} + \gamma_3 \gamma_{vt-1} + \varepsilon_t$$

but with the (non-linear) restriction $\gamma_1\gamma_3 + \gamma_2 = 0$ imposed. This restriction can be tested, but only by using more general methods than we are able to consider here. Of more importance is the implication that residual autocorrelation could be a consequence of omitting important regressors, here x_{t-1} and y_{t-1}.

Modelling the consumption function yet again

As an example of this interpretation of autocorrelation, consider the following dynamic specification of the consumption function

$$\hat{c}_t = \underset{(0.0700)}{0.6778}(y_t - y_{t-1}) + \underset{(0.1239)}{0.3074}\, y_{t-2} + \underset{(0.00041)}{0.00076}(\pi_{t-1} - \pi_{t-2})$$

$$- \underset{(0.00058)}{0.00191}(R_t + R_{t-1})/2 + \underset{(0.0774)}{1.0357}\, c_{t-1} - \underset{(0.0919)}{0.3451}\, c_{t-2} \qquad (14.6)$$

$$R^2 = 0.9996 \qquad \hat{\sigma} = 0.0090$$

This specification has a number of interesting features. Consumption is positively related to the *current growth* of income and the lagged (by two years) *level* of income. One and two year lags of consumption are found to be significant, as well as the lagged *change* in inflation (the acceleration or deceleration of prices) and the average interest rate over the most recent two years. The coefficient estimates of these variables imply that as price rises accelerate real consumption increases, because the cost of a basket of goods will become more expensive if consumption is delayed, while a general increase in interest rates reduces consumption, as saving becomes more attractive.

The presence of the various lags enables distinctions to be made between short-run and long-run effects. For example, the short-run income elasticity is given by the coefficient on y_t, 0.6778, while the long-run elasticity is calculated as the sum of the y coefficients, here 0.3074, divided by one minus the sum of the lagged c coefficients, 0.3094. Hence the long-run elasticity is 0.993 which, interestingly, is significantly less than one.[11]

The moral of this example is that residual autocorrelation is probably more often than not a consequence of *dynamic misspecification*, in the sense that important variables, usually lagged values of already included variables, have been left out of the analysis. This possibility should always be considered before resorting to an 'autocorrelated error' model, as dynamic specifications tend to be much richer and more in accord with the models resulting from economic theorising.[12]

14.8 Other tests of autocorrelation

We have so far discussed two tests for autocorrelation: the Durbin–Watson *dw* statistic and Durbin's *h* test. The former tests for first-order residual autocorrelation, whereas the latter is used if a lagged dependent variable is included as a regressor. These two cases do not, however, exhaust the possible regression specifications that might be encountered. For example, the *h* test is invalid if Y_{t-2} as well as Y_{t-1} is included as a regressor, and it cannot be calculated in some cases (when $TV(\hat{\alpha}) > 1$), while we may also wish to test for residual autocorrelation of a higher order than one, which neither *dw* or *h* is designed for.

We thus need a test that is valid for general dynamic regression specifications of the type encountered above for the consumption function: indeed, is (14.6) actually free of autocorrelation? Such a test is provided by the (Breusch–Godfrey) *Lagrange Multiplier* (LM) test, which is calculated in the following way.[13] Suppose we have regressed Y_t on a set of regressors $X_{1t}, X_{2t}, \ldots, X_{kt}$ which may contain lagged dependent variables (of any order) and lagged independent variables, and obtained the residuals $\hat{u}_t$. If we wish to test for *p*-th order autocorrelation, that is, autocorrelation of the form

$$u_t = \rho_1 u_{t-1} + \rho_2 u_{t-2} + \ldots + \rho_p u_{t-p} + \varepsilon_t$$

we estimate an 'auxiliary' regression of $\hat{u}_t$ on $X_{1t}, X_{2t}, \ldots, X_{kt}$ and $\hat{u}_{t-1}, \hat{u}_{t-2}, \ldots, \hat{u}_{t-p}$ and test the significance of the group $\hat{u}_{t-1}, \hat{u}_{t-2}, \ldots, \hat{u}_{t-p}$ using an *F*-test, which will be distributed as $F(p, T-k-p)$. If annual data is being used, *p* may be set at one or two, say, whereas if the data is quarterly or monthly then *p* may be set at either 4 or 12 respectively, this choice being made so that any *seasonal pattern* in the residuals may be picked up. Of course, any specific value of *p* may be used if that is thought to be appropriate in a specific context.

The LM test statistics for (14.6) with *p* set at 1 and 2 are, respectively, $F(1,54) = 1.15$ and $F(2,53) = 0.56$, which are both insignificant and thus give no indication that the model suffers from residual autocorrelation.

Notes

1 For a brief description of the scope and aims of the subject, see Terence C. Mills, 'Econometrics', in Adam Kuper and Jessica Kuper (editors), *The Social Science Encyclopedia, Volume 1*, 3rd edition (Routledge, 2004), 258–260. G.S. Maddala, *Introduction to Econometrics*, 3rd edition (Wiley, 2001) remains an excellent introductory text.

2 We will not consider assumption 1, that the errors have zero mean, as this will always be satisfied as a consequence of the least squares fit: recall the first

of the normal equations, which says that the sum, and hence the mean, of the residuals is zero. In any case, if the errors did actually have a non-zero mean, this value could simply be added to the intercept.

3 This result may be shown by multiplying (14.2) through by u_{t-j} and taking expectations to obtain

$$\begin{aligned}
E\left(u_t u_{t-j}\right) &= \rho E\left(u_{t-1}u_{t-j}\right) + E\left(\varepsilon_t u_{t-j}\right)\\
&= \rho E\left(\left(\rho u_{t-2} + \varepsilon_{t-1}\right)u_{t-j}\right) = \rho^2 E\left(u_{t-2}u_{t-j}\right) + \rho E\left(\varepsilon_{t-1}u_{t-j}\right) + E\left(\varepsilon_t u_{t-j}\right)\\
&\vdots\\
&= \rho^j E\left(u_{t-j}^2\right) + \rho^{j-1} E\left(\varepsilon_{t-j+1}u_{t-j}\right) + \ldots + E\left(\varepsilon_t u_{t-j}\right)\\
&= \rho^j \sigma^2
\end{aligned}$$

where the results $E\left(u_{t-j}^2\right) = E\left(u_t^2\right) = \sigma^2$ and $E\left(\varepsilon_{t-k}u_{t-j}\right) = 0$ for $k < j$ are used to get the final expression. Now, for $j = 0$,

$$\begin{aligned}
E\left(u_t^2\right) &= \rho E\left(u_{t-1}u_t\right) + E\left(\varepsilon_t u_t\right)\\
&= \rho E\left(u_{t-1}\left(\rho u_{t-1} + \varepsilon_t\right)\right) + E\left(\varepsilon_t\left(\rho u_{t-1} + \varepsilon_t\right)\right)\\
&= \rho^2 E\left(u_{t-1}^2\right) + 2\rho E\left(\varepsilon_t u_{t-1}\right) + E\left(\varepsilon_t^2\right)
\end{aligned}$$

i.e., $\sigma^2 = \rho^2\sigma^2 + \sigma_\varepsilon^2 = \dfrac{\sigma_\varepsilon^2}{1-\rho^2}$

so that

$$E\left(u_t u_{t-j}\right) = \frac{\rho^j \sigma_\varepsilon^2}{1-\rho^2}$$

Note that the assumption that $-1 < \rho < 1$ ensures that the denominator of the expression is positive, so that the covariance depends on the sign of ρ. If ρ is positive then all covariances will be positive, while if ρ is negative the covariances will alternate in sign, being negative for odd j. Furthermore, since $-1 < \rho < 1$ the covariances approach zero as j increases.

4 The proof of this result is as follows:

$$\begin{aligned}
V\left(\hat{\beta}_1\right)_\rho &= \frac{E\left(\sum x_t u_t\right)^2}{\left(\sum x_t^2\right)^2} = \frac{E\left(\sum x_t^2 u_t^2 + 2\sum x_t x_{t-1} u_t u_{t-1} + 2\sum x_t x_{t-2} u_t u_{t-2} + \ldots\right)}{\left(\sum x_t^2\right)^2}\\
&= \frac{\sigma^2}{\left(\sum x_t^2\right)^2}\left(\sum x_t^2 + 2\rho\sum x_t x_{t-1} + 2\rho^2\sum x_t x_{t-2} + \ldots\right)\\
&= \frac{\sigma^2}{\sum x_t^2}\left(1 + 2\rho\frac{\sum x_t x_{t-1}}{\sum x_t^2} + 2\rho^2\frac{\sum x_t x_{t-2}}{\sum x_t^2} + \ldots\right)\\
&= V\left(\hat{\beta}_1\right)\kappa
\end{aligned}$$

5 See Maddala, *op cit.*, p. 240.

6 The spurious regression phenomenon, then referred to as nonsense-correlations, was first pointed out by the famous British statistician G. Udny Yule in the mid-1920s: 'Why do we sometimes get nonsense-correlations between time series? A study in sampling and the nature of time series', *Journal of the Royal Statistical Society* 89 (1926), 1–63. See Terence C. Mills, *The Foundations of Modern Time Series Analysis* (Palgrave Macmillan, 2011), chapter 5, and *A Very British Affair: Six Britons and the Development of Time Series Analysis during the 20th Century* (Palgrave Macmillan, 2013), chapter 2, for extensive discussion.

7 James Durbin and George S. Watson, 'Testing for serial correlation in least squares regression I and II', *Biometrika* 37 (1950), 409–428; 38 (1951), 159–177.

8 Donald Cochrane and Guy H. Orcutt, 'Application of least squares regressions to relationships containing autocorrelated error terms', *Journal of the American Statistical Association* 44 (1949), 32–61.

9 A proof of this assertion is extremely complicated and will not be provided here. It may be found in G.S. Maddala and A.S. Rao, 'Tests for serial correlation in regression models with lagged dependent variables and serially correlated errors', *Econometrica* 41 (1973), 761–774.

10 James Durbin, 'Testing for serial correlation in least squares regression when some of the regressors are lagged dependent variables', *Econometrica* 38 (1970), 410–421.

11 For the dynamic model

$$y_t = \beta_0 x_t + \beta_1 x_{t-1} + \ldots + \beta_r x_{t-r} + \alpha_1 y_{t-1} + \ldots + \alpha_s y_{t-s}$$

the long-run multiplier (or elasticity if y and x are in logs) is given by

$$\frac{\beta_0 + \beta_1 + \ldots + \beta_r}{1 - \alpha_1 - \ldots - \alpha_s}$$

The econometric modelling involved in arriving at the specification (14.6) and tests of the hypotheses contained in this regression are discussed in detail in the *EViews* exercise accompanying this chapter.

12 A technical justification for this position is given by Grayham E. Mizon, 'A simple message for autocorrelation correctors: don't', *Journal of Econometrics* 69 (1995), 267–288.

13 Trevor S. Breusch, 'Testing for autocorrelation in dynamic linear models', *Australian Economic Papers* 17 (1978), 334–355; Leslie G. Godfrey, 'Testing for higher order serial correlation in regression equations when the regressors include lagged dependent variables', *Econometrica* 46 (1978), 1303–1310.

15
Heteroskedasticity

Abstract: *Heteroskedasticity occurs when the error variances are no longer constant across observations, and its presence leads to inefficiency in OLS estimation. Tests for heteroskedaticity are presented, and methods for correcting for its presence are developed. Such methods are often difficult to apply in multiple regression models and an alternative approach is suggested of continuing to use the OLS coefficient estimates but adjusting their standard errors to take into account any heteroskedasticity that might be present. The use of logarithms to mitigate heteroskedasticity is discussed, and an approach to discriminating between linear and logarithmic regression models is proposed.*

15.1 Heteroskedastic errors

Assumption 2 of classical linear regression stated that the variance of the errors was constant across all observations, i.e.,

$$V(u_i) = E(u_i^2) = \sigma^2 \qquad \text{for all } i.$$

This is the assumption of *homoskedasticity*. If it is false, then we have *heteroskedastic* errors. However, non-constancy of the error variance can occur in numerous ways. Let us use the simple bivariate regression for illustration:

$$Y_i = \beta_0 + \beta_1 X_i + u_i \tag{15.1}$$

Note that we have returned to 'i-subscripts': heteroskedasticity can occur in both time series and cross-sectional data. A common example of heteroskedasticity is where the error is a function, not necessarily linear, of the regressor,

$$E(u_i^2) = \sigma_i^2 = \sigma^2 f(X_i) = \sigma^2 Z_i^2 \tag{15.2}$$

Thus if $df/dX_i > 0$, the error variance *increases* as X_i increases, whereas if the derivative is negative, the variance will decrease as the value of the regressor increases. It is conventional to include a common 'scaling factor', σ^2, so that homoskedasticity is a special case of (15.2) when $f(X_i) = 1$. Defining $Z_i^2 = f(X_i)$ enables notation to become a little simpler in what follows. Some forms that commonly occur are $f(X_i) = \delta_0 + \delta_1 X_i^2$, $f(X_i) = \delta_0 + \delta_1 (1/X_i)$ and, if there are two regressors,

$$f(X_{1i}, X_{2i}) = \delta_0 + \delta_1 X_{1i} + \delta_2 X_{2i} + \delta_{11} X_{1i}^2 + \delta_{22} X_{2i}^2 + \delta_{12} X_{1i} X_{2i}$$

so that linear, squared and cross-product terms are all included in the functional form.

Another consumption function

Figure 15.1(a) displays a scatterplot of $N = 20$ observations on consumption expenditure, Y, and income, X, for a cross-sectional sample of families, with the fitted OLS regression line[1]

$$\underset{}{\hat{Y}_i} = \underset{(0.703)}{0.847} + \underset{(0.025)}{0.899}\, X_i \qquad R^2 = 0.986 \qquad RSS = 31.074$$

superimposed.

This looks to provide a good fit to the data, with a very precisely estimated slope of around 0.9, but a less precisely estimated intercept (its *t*-ratio is only 1.2). If we look at a plot of the regression residuals shown in Figure 15.1(b), we get a very clear visual indication of heteroskedasticity in which the error variance increases with income level.

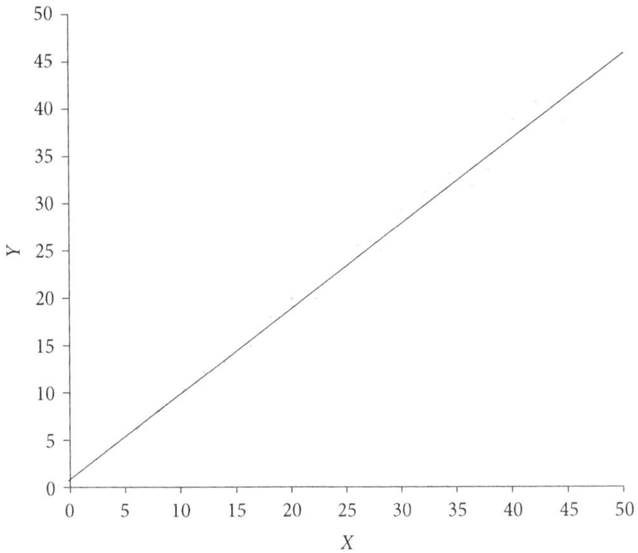

FIGURE 15.1(a) *Heteroskedastic consumption function: consumption–income scatterplot*

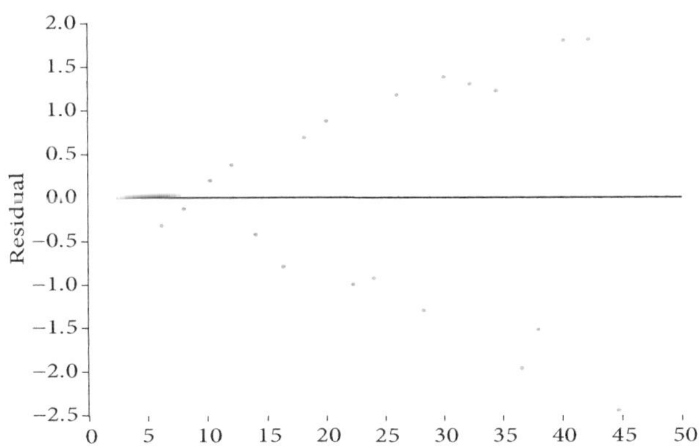

FIGURE 15.1(b) *Heteroskedastic consumption function: consumption–income residuals*

15.2 Consequences of heteroskedasticity

Why should we be concerned with heteroskedastic errors?

We will show that

1. OLS estimators, although remaining unbiased, are *inefficient* in the presence of heteroskedasticity, and
2. the OLS estimated variances of the coefficient estimators are *biased*, thus invalidating, for example, tests of significance.

Consider the regression model given by (15.1) and (15.2). The OLS slope estimator is

$$\hat{\beta}_1 = \frac{\sum xy}{\sum x^2} = \frac{\sum x(\beta_1 x + u)}{\sum x^2} = \beta_1 + \frac{\sum xu}{\sum x^2}$$

Thus $\hat{\beta}_1$ remains unbiased, $E(\hat{\beta}_1) = \beta_1$, since $E\left(\sum xu\right) = 0$ because of assumption 4. However, under heteroskedasticity, the variance of $\hat{\beta}_1$ is[2]

$$V\left(\hat{\beta}_1\right) = V\left(\frac{\sum x_i u_i}{\sum x_i^2}\right) = \frac{1}{\left(\sum x^2\right)^2}\left(x_1^2\sigma_1^2 + \ldots + x_n^2\sigma_n^2\right)$$

which will only collapse to the standard formula $\sigma^2/\sum x^2$ if $\sigma_i^2 = \sigma^2$ for all *i*, i.e., only if we have homoskedastic errors.

Recall the heteroskedastic form (15.2), $\sigma_i^2 = \sigma^2 Z_i^2$, and divide (15.1) by Z_i to obtain

$$\frac{Y_i}{Z_i} = \beta_0 \frac{1}{Z_i} + \beta_1 \frac{X_i}{Z_i} + v_i \qquad (15.3)$$

where the 'new' error, $v_i = u_i/Z_i$, has variance

$$V\left(v_i\right) = E\left(u_i^2/Z_i^2\right) = \sigma^2 Z_i^2/Z_i^2 = \sigma^2$$

that is, 'deflating' by Z_i transforms the error to homoskedasticity.

Because v_i satisfies all the regression assumptions, equation (15.3) can now be estimated by OLS. If we assume $\beta_0 = 0$ for simplicity (or work with the mean deviations *y*, *x* and *z*), this yields

$$\beta_1^* = \frac{\sum (y/z)(x/z)}{\sum (x/z)^2}$$

This, in fact, is the GLS estimator (alternatively known here as the *weighted least squares* (WLS) estimator, because each observation is 'weighted' by Z^{-1}).

β_1^* can be shown to be unbiased and to have variance

$$V(\beta_1^*) = \frac{\sigma^2}{\sum (x/z)^2} \tag{15.4}$$

Noting that we can write

$$V\left(\hat{\beta}_1\right) = \frac{\sum x_i \sigma_i^2}{\left(\sum x_i^2\right)^2} = \frac{\sigma^2 \sum \left(x_i z_i\right)^2}{\left(\sum x_i^2\right)^2}$$

we have that

$$\frac{V\left(\beta_1^*\right)}{V\left(\hat{\beta}_1\right)} = \frac{\sigma^2 \Big/ \sum \left(x/z\right)^2}{\sigma^2 \sum (xz)^2 \Big/ \sum \left(x^2\right)^2} = \frac{\left(\sum x^2\right)^2}{\sum \left(x/z\right)^2 \sum (xz)^2}$$

This ratio will always be less than 1, and will only equal 1 if z is constant. This follows from the fact that ratios of the form $\left(\sum ab\right)^2 \Big/ \sum a^2 \sum b^2$ are always less than or equal to 1, the equality holding if a and b are proportional (recall the Cauchy–Schwarz inequality used in §**5.2**, note 1), that is,

$$V\left(\beta_1^*\right) \le V\left(\hat{\beta}_1\right)$$

and the OLS estimator $\hat{\beta}_1$ is inefficient relative to the GLS estimator β_1^* except when there is no heteroskedasticity, in which case they are identical.

However, this analysis assumes that σ^2, and hence σ_i^2, are known. If, as would typically be the case, we estimate σ^2 from the OLS regression, that is, as $\hat{\sigma}^2 = RSS \big/ (N-2)$, we end up estimating $V\left(\hat{\beta}_1\right)$ by an expression, $\hat{\sigma}^2 \big/ \sum x^2$, whose expected value can be shown to be[3]

$$\frac{\sum x_i^2 \sum \sigma_i^2 - \sum x_i^2 \sigma_i^2}{(N-2)\left(\sum x_i^2\right)^2} \tag{15.5}$$

rather than

$$V\left(\hat{\beta}_1\right) = \frac{\sum x_i^2 \sigma_i^2}{\left(\sum x_i^2\right)^2} \tag{15.6}$$

Under what circumstances will (15.5) be smaller than (15.6)? If

$$\frac{\sum x_i^2 \sum \sigma_i^2 - \sum x_i^2 \sigma_i^2}{(N-2)\left(\sum x_i^2\right)^2} < \frac{\sum x_i^2 \sigma_i^2}{\left(\sum x_i^2\right)^2}$$

then

$$\frac{1}{N-1} < \frac{\sum x_i^2 \sigma_i^2}{\sum x_i^2 \sum \sigma_i^2}$$

which will be guaranteed if $\sum x_i^2 \sigma_i^2 > 0$, i.e., if σ_i^2 and x_i^2 are positively correlated, as is often the case.

In these circumstances the true variance of the OLS estimator is *underestimated* and so, for example, we get shorter confidence intervals and larger test statistics than we should do: we suffer from 'false precision'. It is important to emphasise this point. Although the *true* OLS coefficient variance (15.6) (which takes into account heteroskedasticity) is *larger* than the optimal GLS coefficient variance (15.4), OLS regression, which ignores heteroskedasticity, will use the *wrong formula* (15.5) which underestimates (15.6) (cf. the situation when dealing with autocorrelation in **§14.2**).

15.3 A consistent variance estimate

Although the OLS formula for the variance of $\hat{\beta}_1$ is incorrect, being both biased and inconsistent, we could use an estimator of (15.6) to obtain a better estimate of the variance. A consistent estimator of the changing variance σ_i^2 is, in fact, provided by the *squared residual* from the OLS regression, $\hat{u}_i^2$. Substituting this into (15.6) yields *White's consistent variance estimator*[4]

$$\hat{V}\left(\hat{\beta}_1\right) = \frac{\sum x_i^2 \hat{u}_i^2}{\left(\sum x_i^2\right)^2}$$

This is useful when the form of Z_i is not completely known, as may well be the case when there are several regressors. For our consumption data, using White's estimator gives a standard error on $\hat{\beta}_1$ of 0.028, slightly larger than the OLS standard error.

15.4 Testing for heteroskedasticity

Although there are a variety of tests for heteroskedasticity, we shall only concentrate on two, as these are the easiest to compute and interpret. The first is *White's*, and follows the same logic as the consistent variance estimator in §15.3, where it was stated that $\hat{u}_i^2$ was a consistent estimator of σ_i^2. Thus, recalling (5.2), a general test of heteroskedasticity would be to regress $\hat{u}_i^2$ on a function of the regressor(s). The function chosen is the polynomial with cross-products, as this will be a decent approximation to most functional forms; for example, if we have two regressors, the *auxiliary* regression

$$\hat{u}_i^2 = \delta_0 + \delta_1 X_{1i} + \delta_2 X_{2i} + \delta_{11} X_{1i}^2 + \delta_{22} X_{2i}^2 + \delta_{12} X_{1i} X_{2i} + e_i$$

is fitted and the null hypothesis of homoskedasticity, here

$$H_0 : \delta_1 = \delta_2 = \delta_{11} = \delta_{12} = \delta_{22} = 0$$

is tested, most straightforwardly using a standard *F*-test,

$$F_W = \frac{R_A^2}{1 - R_A^2} \frac{N - m - 1}{m} \sim F(m, N - m - 1)$$

where R_A^2 is the R^2 from the auxiliary regression and m is the number of regressors in that regression: $m = 2$ if $k = 1$, $m = 5$ if $k = 2$, $m = 9$ if $k = 3$, etc.

The second test is that of *Goldfeld and Quandt*, which is useful when it is felt that the error variance is either increasing or decreasing with the values of a particular regressor.[5] Suppose we have a single regressor and we reorder the data in ascending values of X_i. The data is then split into two groups of N_1 and N_2 observations, corresponding to small and large values of X ($N_1 + N_2$ can be less than N, as observations may be omitted from the 'middle' of the data to improve the performance of the test). Separate regressions are then run, obtaining error variance estimates $\hat{\sigma}_1^2$ and $\hat{\sigma}_2^2$. The statistic

$$F_{GQ} = \frac{\hat{\sigma}_2^2}{\hat{\sigma}_1^2} \sim F(N_1 - 2, N_2 - 2)$$

is then calculated, and the null of homoskedasticity, H_0: $\sigma_1^2 = \sigma_2^2$, is rejected for significantly large values of F_{GQ}. (This assumes that the variance is increasing with X, so that $\hat{\sigma}_1^2 < \hat{\sigma}_2^2$. If the converse occurs, the terms in the F-ratio may be reversed, thus ensuring that it always exceeds 1.)

Testing the consumption function for heteroskedasticity

Using the residuals from the OLS regression, the following auxiliary regression was estimated

$$\hat{u}_i^2 = \underset{(0.618)}{0.488} - \underset{(0.055)}{0.070}\, X_i + \underset{(0.0011)}{0.0037}\, X_i^2 \qquad R_A^2 = 0.878$$

from which we calculate $F_W = 61.2 \sim F(2,17)$, which is clearly significant. Since $d\hat{u}_i^2/dX_i = -0.070 + 0.0074X_i$, only for the two smallest values of X, $X_1 = 6.2$ and $X_2 = 8.1$, is the slope of the relationship (just) negative: otherwise it is positive and increasing, as is seen in the plot of $\hat{u}^2$ on X shown in Figure 15.2.

To compute the Goldfeld–Quandt test, the sample may be split in half, so that $N_1 = N_2 = 10$, the small sample size precluding any observations from the middle being omitted. The two regressions are estimated as

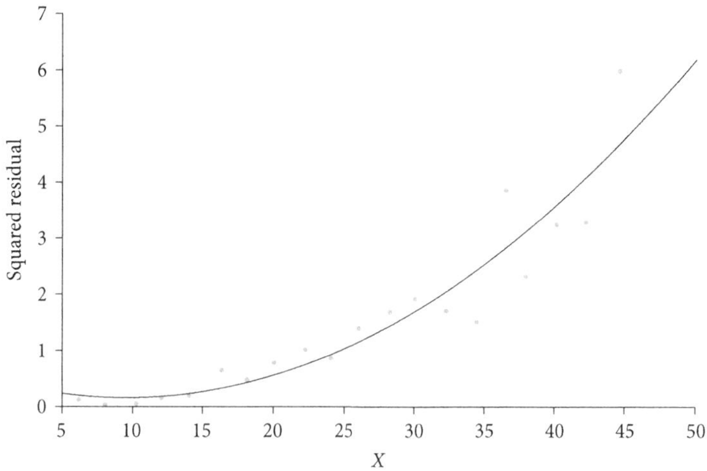

FIGURE 15.2 *Scatterplot of squared residuals on X*

$$\hat{Y}_i = 1.053 + \underset{(0.616)}{} \; 0.876 X_i \qquad R_1^2 = 0.985 \qquad \hat{\sigma}_1^2 = 0.475$$
$$\qquad (0.616) \quad (0.038)$$

$$\hat{Y}_i = 3.279 + 0.835 X_i \qquad R_2^2 = 0.904 \qquad \hat{\sigma}_2^2 = 3.154$$
$$\qquad (3.443) \quad (0.096)$$

and hence

$$F_{GQ} = \frac{3.154}{0.475} = 6.64 \sim F(8,8)$$

which again is significant ($F_{0.01}(8,8) = 6.03$). Both tests thus reject (not surprisingly) the null of homoskedasticity and although the alternative hypotheses are different, both imply that larger values of X are associated with larger error variances.

15.5 Correcting for heteroskedasticity

We have already seen how to correct for heteroskedasticity by using GLS. If we have the model

$$Y_i = \beta_0 + \beta_1 X_i + u_i$$

and it is known that

$$E\left(u_i^2\right) = \sigma^2 Z_i^2$$

then we can divide through by Z and use OLS on the transformed model

$$\frac{Y_i}{Z_i} = \beta_0 \frac{1}{Z_i} + \beta_1 \frac{X_i}{Z_i} + v_i$$

where

$$E\left(v_i^2\right) = E\left(u_i / Z_i\right)^2 = \sigma^2$$

However, it is important to note that if there is an intercept in the original equation, then there *will not be one* in the transformed equation: Y/Z is regressed on $1/Z$ and X/Z alone, and the estimate of the slope coefficient on the regressor $1/Z$ is the estimate of the intercept.

Of course, in practice Z will be unknown, and a simple choice is to set it equal to the regressor X. Another possibility is to use the auxiliary regression from the White test:

$$Z_i^2 = 0.488 - 0.070X_i + 0.0037X_i^2$$

Thus, using $Z_i = \left(0.488 - 0.070X_i + 0.0037X_i^2\right)^{\frac{1}{2}}$, we obtain

$$\frac{\hat{Y}_i}{Z_i} = \underset{(0.330)}{0.727}\frac{1}{Z_i} + \underset{(0.020)}{0.905}\frac{X_i}{Z_i}, \qquad R^2 = 0.929$$

Recalling that the OLS estimates were $\hat{\beta}_0 = 0.847(0.703)$ and $\hat{\beta}_1 = 0.899(0.025)$, we see that the slope is estimated to be a little larger and is a little more precisely estimated, while the intercept, although numerically smaller, is now established to be reliably positive, with a t-ratio of 2.2.

It is difficult to provide rules for selecting Zs, and the problem is compounded when dealing with multiple regressions. Indeed, the most popular approach currently is to simply rely on the OLS coefficient estimates but to report the White standard errors (the White standard error for the intercept of the consumption function is 0.527, leading to a t-ratio of 1.61).

15.6 The logarithmic transformation again

It is often the case that transforming the variables to logarithms will also transform a heteroskedastic error to a homoskedastic one, since logging variables compresses the scales on which they are measured (recall §3.2).

Does taking logarithms work for the consumption regression? Here the evidence is somewhat mixed, as the White test still rejects homoskedasticity ($F_W = 13.1$) but the Goldfeld–Quandt test does not (F_{GQ} is only 1.75).

The logarithmic regression is

$$\ln\hat{Y}_i = \underset{(0.0574)}{0.0757} + \underset{(0.018)}{0.956}\ln X_i \qquad R^2 = 0.9935$$

Again, R^2s and σ^2s cannot be compared across logarithmic and linear regressions, but the following decision rule may be used. Define $\hat{\sigma}^2$ and $\tilde{\sigma}^2$ to be the error variances from the linear and log–linear regressions respectively, and calculate

$$L_{\text{lin}} = -\frac{N}{2}\ln\hat{\sigma}^2 \qquad \text{and} \qquad L_{\log} = -\frac{N}{2}\ln\tilde{\sigma}^2 - \sum \ln Y_i$$

The second term in the $L_{\log}$ expression is there to correct for the different units of measurement. We then select the model that yields the largest L value. For the linear regression, we have $\hat{\sigma} = 1.3139$ and $L_{\text{lin}} = -5.460$ while for the log–linear model we have $\tilde{\sigma} = 0.0457$ and $L_{\log} = 1.035$. We therefore select the log–linear model as the most appropriate functional form.

The two models imply different values for the income elasticity of consumption. The log–linear regression gives a *constant* elasticity of 0.956, while from the linear regression the elasticity varies as

$$\eta_i = 0.899\frac{X_i}{Y_i}$$

which at the mean values of consumption and income is computed as

$$\bar{\eta} = 0.899\frac{25.25}{23.56} = 0.963$$

From these two regressions, can we infer that the elasticity is significantly less than unity? A 95% confidence interval for the elasticity obtained directly from the logarithmic regression is $0.918 < \eta < 0.995$, while a similar calculation from the linear regression might take the form $0.840(25.25/23.56) < \bar{\eta} < 0.959(25.25/23.56)$, that is, $0.900 < \bar{\eta} < 1.028$, although this calculation takes no account of any sampling error in computing the means.

Notes

1 Data taken from G.S. Maddala, *Introduction to Econometrics*, 3rd edition (Wiley, 2001), table 5.1.

2 A derivation of this result is as follows.

$$V(\hat{\beta}_1) = E(\hat{\beta}_1 - \beta_1)^2 = E\left(\frac{\sum xu}{\sum x^2}\right)^2 = \frac{E\left(\sum xu\right)^2}{\left(\sum x^2\right)^2}$$

$$= \frac{E(x_1u_1 + x_2u_2 + \ldots + x_Nu_N)(x_1u_1 + x_2u_2 + \ldots + x_Nu_N)}{\left(\sum x^2\right)^2}$$

$$= \frac{E\left(x_1^2 u_1^2 + x_2^2 u_2^2 + \ldots + x_N^2 u_N^2\right)}{\left(\sum x^2\right)^2}$$

$$= \frac{x_1^2 \sigma_1^2 + x_2^2 \sigma_2^2 + \ldots + x_N^2 \sigma_N^2}{\left(\sum x^2\right)^2}$$

where we make use of assumption 3, so that $E(x_i x_j u_i u_j) = 0$ whenever $i \neq j$.

3 To show this result we must first obtain an expression for the expected value of the residual sum of squares:

$$E(RSS) = E\left(\sum\left(y - \hat{\beta}_1 x\right)^2\right) = E\left(\sum y^2 + \hat{\beta}_1^2 \sum x^2 - 2\hat{\beta}_1 \sum xy\right)$$

$$= E\left(\sum(\beta_1 x + u)^2 + \left(\beta_1 + \frac{\sum xu}{\sum x^2}\right)^2 \sum x^2 - 2\left(\beta_1 + \frac{\sum xu}{\sum x^2}\right)\left(\beta_1 \sum x^2 + \sum xu\right)\right)$$

$$= E\left(\begin{array}{l} \beta_1^2 \sum x^2 + \sum u^2 + 2\beta_1 \sum xu + \beta_1^2 \sum x^2 + \dfrac{\left(\sum xu\right)^2}{\sum x^2} + 2\beta_1 \sum xu \\ -2\beta_1^2 \sum x^2 - 4\beta_1 \sum xu - 2\dfrac{\left(\sum xu\right)^2}{\sum x^2} \end{array}\right)$$

$$= E\left(\sum u^2 - \frac{\left(\sum xu\right)^2}{\sum x^2}\right)$$

$$= \sum \sigma_i^2 - \frac{\sum x_i^2 \sigma_i^2}{\sum x_i^2}$$

Thus

$$E\left(\hat{\sigma}^2/(N-2)\right) = \frac{E(RSS)}{(N-2)\sum x_i^2} = \frac{\sum x_i^2 \sum \sigma_i^2 - \sum x_i^2 \sigma_i^2}{(N-2)\sum x_i^2}$$

4 Halbert White, 'A heteroskedasticity consistent covariance matrix estimator and a direct test of heteroskedasticity', *Econometrica* 48 (1980), 817–838.

5 Steven M. Goldfeld and Richard E. Quandt, 'Some tests for homoskedasticity', *Journal of the American Statistical Association* 60 (1965), 539–547.

16

Simultaneity, Instrumental Variables and Non-Normal Errors

Abstract: *Simultaneity occurs when the regressor and the error are no longer independent, as is required in the classical assumptions. This leads to simultaneity bias, while other violations of this assumption, which can occur regularly with economic data, also lead to biased estimates, in particular when autocorrelation and a lagged dependent variable appear together. Instrumental variables estimation is a potential solution to this problem. The presence of non-normally distributed errors may be symptomatic of important misspecifications that could easily lead to very misleading and, in some cases, completely ridiculous coefficient estimates. Residuals should be examined for the presence of outliers and a test for non-normality is presented, with examples being used to illustrate the impact and mitigation of outlying observations.*

16.1 Stochastic regressors

In this chapter we examine the effects of violations to the classical regression assumptions 4 and 5 of §12.2. Assumption 4 is that the errors and the regressor are independent. Strictly, the results derived in §12.2 assume that the regressor is non-stochastic, so that $E(u_i x_j) = x_j E(u_i) = 0$ using assumption 1. This is clearly unrealistic in most economic situations, so when it is assumed that the regressor is stochastic but independent of the error, how valid are the results? Unbiasedness continues to hold, and the formulae for estimators, variances and covariances – and hence test statistics derived from them – continue to be correct if y and x are jointly normally distributed. If this cannot be assumed, then these have to be viewed as being conditional on the observed values of x.[1]

Violations of the assumption of independence between the regressors and the errors may, however, occur in various situations, and will typically lead to biased estimates. An appropriate method of estimation in these circumstances is known as *instrumental variables.*

16.2 Simultaneity bias

Assumption 4 of classical linear regression states that

$$Cov(u_i, x_j) = E(u_i x_j) = 0$$

for all i and j.

If this assumption is violated then, using the simple bivariate regression model as an example,

$$E(\hat{\beta}_1) = \beta_1 + \frac{E\left(\sum xu\right)}{\sum x^2} \neq \beta_1 \tag{16.1}$$

because we cannot set $E(xu)$ to zero, so that the OLS estimator is then *biased.*

One commonly encountered situation in which this assumption is false is that of *simultaneity bias*, which we can illustrate using the simple (time series) consumption function

$$c_t = \beta y_t + u_t \tag{16.2}$$

So far we have considered the consumption function in isolation from the rest of the macroeconomy, but we now recall from §5.4 that consumption and income are also linked in the Keynesian model through the national income accounting identity, which in its simplest form is

$$y_t = c_t + i_t \tag{16.3}$$

where i_t is investment. Now that we have the 'system' of equations (16.2) and (16.3), what are the properties of the OLS estimator $\hat{\beta}$ from the regression of (16.2)? Substituting (16.2) into (16.3) gives

$$y_t = \frac{i_t}{1-\beta} + \frac{u_t}{1-\beta} = \frac{i_t + u_t}{1-\beta}$$

so that

$$E(y_t u_t) = E\left(\frac{(i_t + u_t)u_t}{1-\beta}\right) = E\left(\frac{i_t u_t + u_t^2}{1-\beta}\right) = E\left(\frac{u_t^2}{1-\beta}\right) = \frac{\sigma^2}{1-\beta}$$

on the assumption that $E(i_t u_t) = 0$, that is, that the errors in the consumption function are independent of investment, so that i_t is *exogenous*. Now, using (16.1), we have

$$E(\hat{\beta}) = \beta + \frac{E\left(\sum y_t u_t / T\right)}{E\left(\sum y_t^2 / T\right)} = \beta + \frac{\sigma^2 / (1-\beta)}{\sigma_y^2} = \beta + \frac{1}{1-\beta}\left(\frac{\sigma^2}{\sigma_y^2}\right)$$

Here we have used the sample size T to 'scale' the two expectations (which contain sums of T terms) and have taken account of the fact that the presence of the national accounting identity effectively makes y a random variable, so that it has variance $\sigma_y^2 = E\left(\sum y^2 / T\right)$.

If $0 < \beta < 1$, as would be suggested by the theory of the consumption function, it follows that the bias $E(\hat{\beta}) - \beta$ must be *positive*, so that the OLS estimator $\hat{\beta}$ *overestimates* β. For example, in the linear consumption function used as an example in Chapter 15, $\hat{\sigma} = 1.3139$ and $\sigma_y = 11.910$, so that the bias is

$$\frac{1}{1-\beta} \cdot 0.0122$$

Thus if the true β is 0.80, the bias is 0.06 and the expected value of $\hat{\beta}$ is 0.86 (recall that the actual estimate was 0.90). As a second example, the log–linear consumption function of §**12.4** has $\hat{\sigma}=0.0257$ and $s_y=0.4451$, so that

$$E(\hat{\beta})=\beta+\frac{1}{1-\beta}\cdot 0.0033$$

Substituting $\hat{\beta}=0.982$ for the expectation, solving the resultant quadratic in β and taking the 'economic' solution, gives $\beta=0.938$; that is, an OLS estimate of 0.982 for β implies an unbiased estimate of 0.938.

16.3 Errors in variables

A second way in which assumption 4 may be violated is when the variables are *measured with error*. By this we do not mean that the values taken by the variables have been recorded incorrectly: we will discuss this possibility later (see §**16.6**). Rather, we are looking at the case when the errors are due to using an imperfect measure of the true variable. Such imperfect measures are known as *proxy variables*, while the true variables, which are often not measurable, are called *latent variables*. Examples might be expectational variables, which must be replaced by proxies (either estimated from past observations or obtained from survey data), variables which do not match the definition required by economic theory (one important example uses measured income as a proxy for the theoretically more appropriate concept of *permanent income* in the consumption function), and the use of 'agency', rather than economic, definitions of a variable (for example, the rate of inflation of the RPI may not correspond to the rate of inflation felt by any particular individual or group).

Thus suppose that the *true* model is

$$y^* = \beta x^* + e \tag{16.4}$$

but, instead of y^* and x^*, we measure

$$y = y^* + v \quad \text{and } x = x^* + u$$

u and v are measurement errors, which we will assume to have zero means and variances σ_u^2 and σ_v^2, respectively. We will also assume that they are mutually uncorrelated and also uncorrelated with the latent variables:

$$E(uv) = E(ux^*) = E(uy^*) = E(vx^*) = E(vy^*) = 0 \tag{16.5}$$

Equation (16.4) can be written as

$$y - v = \beta(x - u) + e$$

or

$$y = \beta x + w \tag{16.6}$$

where the error term is a linear combination of the three individual errors

$$w = e + v - \beta u$$

Now

$$E(wx) = E\left(\left(e + v - \beta u\right)\left(x^* + u\right)\right) = -\beta E\left(u^2\right) = -\beta\sigma_u^2$$

using the conditions (16.5). This is clearly non-zero, thus violating assumption 4. Note, however, that $E(wx)$ does not depend upon σ_v^2, so that measurement error in *y alone* does not cause the problem: this is due solely to measurement error in *x*.

It can be shown that the bias in the OLS estimator $\hat{\beta}$ is given by[2]

$$\beta\left(\frac{\sigma_{x^*}^2}{\sigma_{x^*}^2 + \sigma_u^2} - 1\right) \tag{16.7}$$

Since the term in brackets must be negative, the OLS estimator is *biased downwards*. Thus, if consumption is really dependent upon permanent income, but measured income is used instead, we obtain *too small* an estimate of β.

16.4 Lagged dependent variables and autocorrelation

A third way in which assumption 4 may be violated is when the model contains *both* a lagged dependent variable *and* an autocorrelated error:

$$y_t = \beta x_t + \alpha y_{t-1} + u_t$$
$$u_t = \rho u_{t-1} + \varepsilon_t$$

In the absence of autocorrelation, OLS provides unbiased (although inefficient: recall §14.3) estimates of α and β. However, with an autocorrelated error we now have

$$\begin{aligned} E\left(u_t y_{t-1}\right) &= E\left(\left(\rho u_{t-1} + \varepsilon_t\right)\left(\beta x_{t-1} + \alpha y_{t-2} + u_{t-1}\right)\right) \\ &= \rho E\left(u_{t-1}^2\right) \\ &= \rho\sigma^2 \neq 0 \end{aligned}$$

This non-zero expectation ensures that *both* $\hat{\alpha}$ and $\hat{\beta}$ are *biased*. It can also be shown that if $\rho > 0$ then $\hat{\alpha}$ is biased upwards by the same amount that the estimate of the autocorrelation parameter, $\hat{\rho}$, is biased downwards. This is the reason why $dw = 2(1 - \hat{\rho})$ is biased towards 2 in the presence of a lagged dependent variable.

16.5 Instrumental variables estimation

An appropriate estimation technique when assumption 4 is violated is that of *instrumental variables* (IV). The reason why we cannot use OLS to estimate (16.6) is that the error w is correlated with the regressor x. The IV method consists of finding a variable z that is *uncorrelated* with w but *correlated* with x and then estimating β by

$$\hat{\beta}_{IV} = \frac{\sum yz}{\sum xz} \tag{16.8}$$

The variable z is called an 'instrumental variable'. Note that the OLS estimator may be regarded as an IV estimator where the instrumental variable is x itself. Why does this method 'work', in the sense of producing a consistent estimator of β? We can write the IV estimator as

$$\hat{\beta}_{IV} = \frac{\sum(\beta x + w)z}{\sum xz} = \beta + \frac{\sum wz}{\sum xz}$$

The condition that the instrument z is uncorrelated with the error w but correlated with the regressor x ensures that the expectation of the ratio $\Sigma wz/\Sigma xz$ is zero, thus showing that $\hat{\beta}_{IV}$ is unbiased (strictly, consistent).

Where do we get these instrumental variables from? They will usually be suggested by the particular problem under investigation. For example, in the consumption function, investment is by definition correlated with income, but since it is assumed to be exogenous it is also uncorrelated with the error in (16.2). It thus satisfies both conditions for an IV, so that a consistent estimate of β is given by (and noting the unfortunate change in definition of y compared to (16.8)!)

$$\hat{\beta}_{IV} = \frac{\sum ci}{\sum yi}$$

When we have a lagged dependent variable and an autocorrelated error (as in §16.4), we typically use the *lagged independent variable*, x_{t-1}, as the instrument. Again, this is uncorrelated with the error by definition, but will certainly be correlated with y_{t-1} by the nature of the model.

16.6 Non-normal errors

Assumption 5 of classical linear regression concerns the distribution of the errors; it was assumed that they were *normally* distributed: $u \sim IN(0,\sigma^2)$. This is the 'forgotten assumption' of regression, and is often completely ignored in many applications. This neglect has typically been justified on the grounds that the OLS estimator remains BLUE under non-normal errors, and as long as the sample is large enough, hypothesis tests based on a normal assumption will still be appropriate.

However, non-normal residuals are often symptomatic of important misspecifications that might lead to the actual estimate of $\hat{\beta}$ being misleading, and in some cases completely ridiculous! The easiest way to detect non-normality is to examine a plot or the histogram of the residuals. This will often 'flag' one or two outliers in the data that may have an undue influence on the regression fit. These may be due to exogenous shocks, a strike or change in definition, for example, or may even be caused by a recording error. Typing in an incorrect number – a decimal point placed wrongly, say – can have an amazing affect on the fitted line.

A formal test of normality is available. This is the *Jarque–Bera* statistic[3]

$$JB = \frac{n}{6}\hat{\alpha}_3^2 + \frac{n}{24}\left(\hat{\alpha}_4 - 3\right)^2$$

Here $\hat{\alpha}_3$ is a measure of the amount of skewness in the residuals (recall §2.5), while $\hat{\alpha}_4$ measures the amount of kurtosis (the extent to which the error distribution deviates from the 'bell' shape of the normal distribution). For a normal distribution, which is symmetric, skewness is obviously zero, and it has a kurtosis of 3. Thus $\hat{\alpha}_4 - 3$ measures the *excess kurtosis* in the residuals: if kurtosis is greater than 3 then the error distribution is said to be *fat-tailed* and will have more outliers than it should have if it was normally distributed. This is the typical occurrence, and only very occasionally do we find kurtosis to be significantly less than 3 and the distribution to be thin-tailed (the uniform distribution introduced in §8.2 is an example of such a distribution). The *JB* statistic thus jointly tests for zero skewness and excess kurtosis, and under the null of normality should be zero.

The statistic is, in fact, asymptotically distributed as $\chi^2(2)$ so that, since $\chi^2_{0.05}(2) = 5.99$, a simple decision rule is available: reject the null of normality if $JB \geq 6$. In fact, if there is non-normality, *JB* will often take values vastly in excess of 6. Some care should be taken with small sample sizes, however, as the *JB* statistic is then overly sensitive, having too large a Type I error probability: for example, using the above decision rule when $T = 30$ will lead to an incorrect rejection probability in the region of 10% rather than 5%.

Two examples of non-normal errors

As an example of how a 'data transcription' error can affect OLS regression, let us return to our original consumption function example first introduced as Figure 5.2. The 1995 observation on consumption was 832450, that is, $C_{48} = 832450$. Suppose that when entering the data via a keyboard we inadvertently typed this value as 8324500. The resultant scatterplot is shown in Figure 16.1, where it is apparent that the single outlier clearly 'drags' the regression line towards it (cf. the outlier example in §5.3).

The estimated line is

$$\underset{}{\hat{C}_t} = \underset{(295816)}{-66865} + \underset{(0.35)}{1.10}\, Y_t \qquad R^2 = 0.136$$

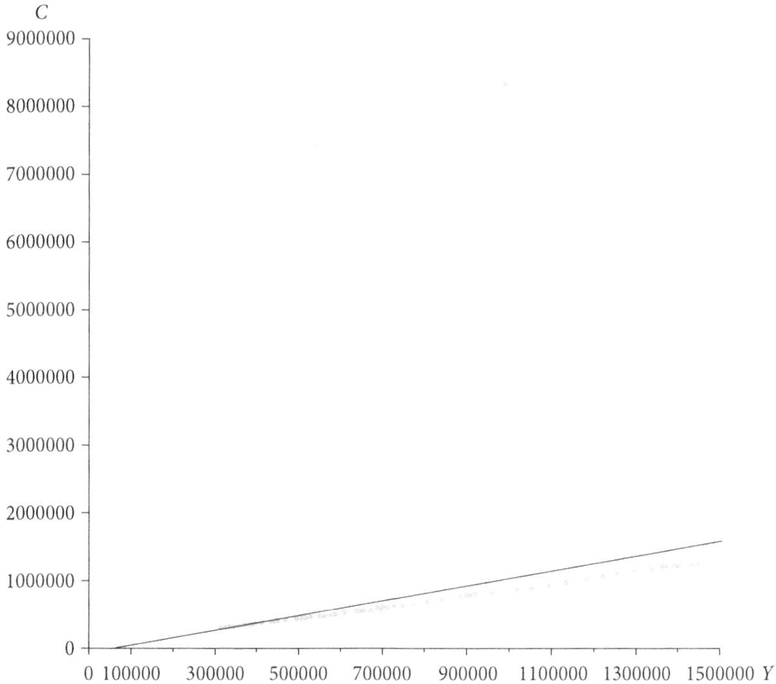

FIGURE 16.1 *The effect of a data transcription error*

Clearly, with $\hat{\beta}_0 < 0$ and $\hat{\beta}_1 > 1$, something is seriously amiss! Obviously, a look at the scatterplot shows the problem immediately, and this should *always* be done prior to any regression analysis.

Nonetheless, the Jarque–Bera statistic does indeed reveal a serious problem of non-normality, for it takes the value $JB = 9177.3$. Moreover, $\hat{\alpha}_3 = 7.7$ and $\hat{\alpha}_4 = 60.1$, so that the distribution of the residuals is both highly skewed and fat-tailed – and all this because a single data value has been mistyped!

The solution here is obvious: re-enter C_{48} as the correct value! Often, however, large outliers occur naturally, rather than artificially, and a simple technique usually accounts for their influence. This 'trick' can be used here as well, and employs the *dummy variable* defined as

$$D_{48,t} = \begin{cases} 1 & \text{if} \quad t = 48 \\ 0 & \text{if} \quad t \neq 48 \end{cases}$$

This is included as an additional regressor, to obtain

$$\hat{C}_t = -\ 4837 +\ 0.861\ Y_t + 7476501 D_{48,t}$$
$$\quad (4321)\ (0.005)\quad (13980)$$
$$R^2 = 0.9998;\ \hat{\alpha}_3 = -0.14;\ \hat{\alpha}_4 = 2.14;\ JB = 2.14$$

This technique 'removes' the outlier in the sense that the predictions for C_t are

$$\hat{C}_{48} = -4837 + 0.861 Y_{48} + 7476501,$$
$$\hat{C}_t = -4837 + 0.861 Y_i \qquad t \neq 48$$

and no evidence of non-normality remains in the residuals.

Note, however, that the estimates from this regression are *not quite* identical to those from the regression on the 'correct' data, which is $C_t = -4798 + 0.860 Y_t$. This is because the presence of the dummy forces the residual for the data point (C_{48}, Y_{48}) to be *exactly* zero, which it will not be for the 'correct' regression (where it is actually $\hat{u}_{48} = -15191.5$), and this leads to the slight differences in the estimates.

As a second example, consider the regression of the change in the long UK interest rate (the yield on 20-year gilt edged stock, ΔR_t) on the change in the short UK interest rate (the yield on three-month Treasury bills, Δr_t) estimated on monthly observations from March 1952 to June 2012. A scatterplot of the data with the fitted regression line $\Delta R_t = -0.002 + 0.307 \Delta r_t + \hat{u}_t$ superimposed is shown in Figure 16.2, while the histogram of the residuals $\hat{u}_t$, with a normal distribution with the same mean and variance as the residuals superimposed, is shown in Figure 16.3.

It is clear from these figures that the residuals are highly non-normal, and this is confirmed by the statistics $\hat{\alpha}_3 = -0.3$, $\hat{\alpha}_4 = 5.6$ and $JB = 222.5$. In this case there are far too many outliers for each to be modelled by a dummy variable, and we must conclude that the error term is simply not normally distributed. This conclusion implies that estimation methods other than OLS must be used, but these are too advanced to be discussed here.

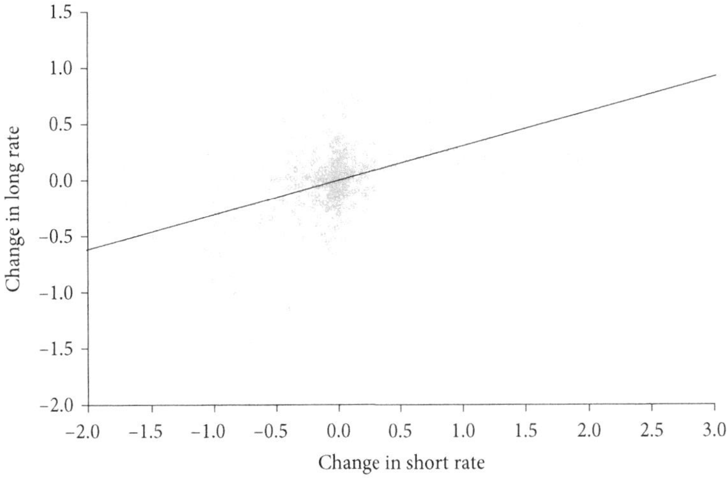

FIGURE 16.2 *Scatterplot of ΔR_t and Δr_t with regression line superimposed*

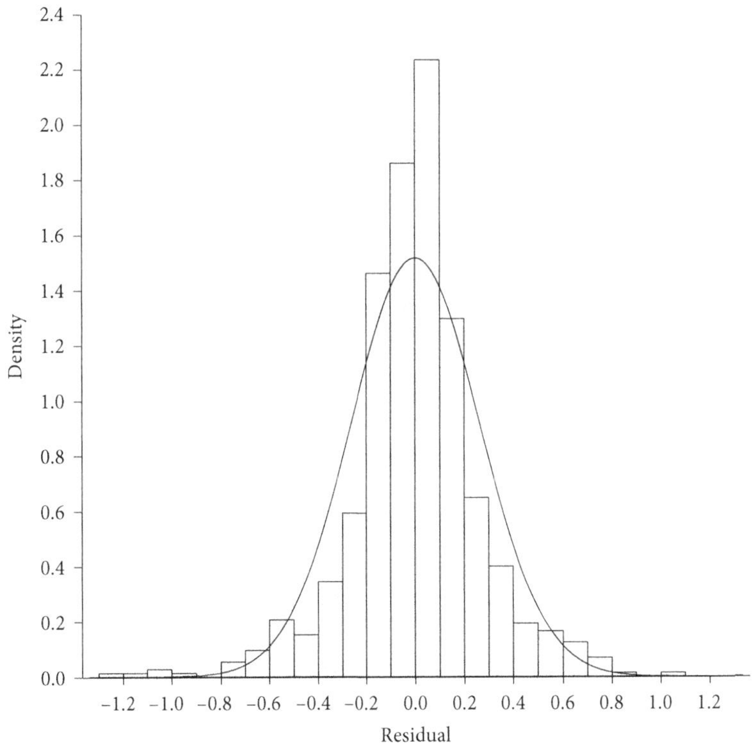

FIGURE 16.3 *Histogram of residuals with normal distribution superimposed*

Notes

1 It is straightforward to show unbiasedness under the assumption of independence using the expression

$$\hat{\beta}_1 = \beta_1 + \frac{\sum xu}{\sum x^2}$$

On taking expectations under independence we have

$$E\left(\hat{\beta}_1\right) = \beta_1 + E\left(\frac{\sum xu}{\sum x^2}\right) = \beta_1 + \sum E\left(\frac{x}{\sum x^2}\right) E\left(u\right) = \beta_1$$

using assumption 1 and without having to evaluate $E\left(x/\sum x^2\right)$. The results for the variances and covariances under normality are much more difficult to obtain, but are conveniently summarised in Allan R. Sampson, 'A tale of two regressions', *Journal of the American Statistical Association* 69 (1974), 682–689.

2 The OLS estimator of β in (16.6) is

$$\hat{\beta} = \frac{\sum xy}{\sum x^2} = \frac{\sum\left(x^* + u\right)\left(y^* + v\right)}{\sum\left(x^* + u\right)^2}$$

Multiplying $y^* = \beta x^* + e$ by x^* and taking expectations yields $\beta = \sigma_{x^*y^*} / \sigma^2_{x^*}$, where $\sigma_{x^*y^*} = E\left(x^* y^*\right) = Cov\left(x^*, y^*\right)$. Hence, on taking the expectation of $\hat{\beta}$ and using the conditions (16.5),

$$E\left(\hat{\beta}\right) = \frac{\sigma_{x^*y^*}}{\sigma^2_{x^*} + \sigma^2_u} = \frac{\beta\sigma^2_{x^*}}{\sigma^2_{x^*} + \sigma^2_u}$$

and it then follows that $E(\hat{\beta}) - \beta$ is given by (16.7).

3 Carlos M. Jarque and Anil K. Bera, 'Efficient tests for normality, homoscedasticity and serial dependence in regression residuals', *Economics Letters* 6 (1980), 255–259.

17
Testing for Stability in Regression Models

Abstract: *An implicit assumption in all regression models is that their coefficients remain constant across all observations. When they do not – and this occurs regularly with time series data in particular – the problem of structural change is encountered. After presenting a simulation example of a typical structural break in a regression, methods are introduced to test for such breaks, whether at a known point in time or when the break-point is unknown. An approach to modelling changing parameters using dummy variables is introduced and a detailed example of a shifting regression relationship between inflation and interest rates brought about by policy regime changes is presented.*

17.1 Structural stability in regression models

An implicit assumption in all regression models is that their coefficients remain constant across all observations. Particularly with time series data, this may not necessarily be the case, since coefficients can 'evolve' through time, or indeed alter abruptly at a point in time, in many different ways. This is known as the problem of *structural change*, which has a tendency to pervade many attempts at modelling economic data.

A simple, but plausible, artificial example will be used to demonstrate the problems that may arise with structurally changing models. Consider a (static) consumption function, $C_t = \beta_0 + \beta_1 Y_t + u_t$, estimated using annual data for the 20th century, that is, we have $t = 1900, 1901, \ldots, 1999$. It would not be surprising if both the average propensity to consume, $APC_t = C_t/Y_t = \beta_1 + (\beta_0/Y_t)$, and the marginal propensity to consume, $MPC = \beta_1$, were smaller during war years than during peacetime, as expenditure is switched away from consumption into war production. This would imply that both the intercept, β_0, and the slope, β_1, take different values during war time to the values they take during peacetime.

Suppose that the consumption function took the following form during peacetime

$$C_t = 10 + 0.8Y_t + u_t\,,\quad u_t \sim IN(0,100)$$

$$t = 1900,\ldots,1913,\ 1919,\ldots,1938,\ 1946,\ldots,1999$$

and the form

$$C_t = 6 + 0.5Y_t + u_t,\quad u_t \sim IN\left(0,100\right)$$

$$t = 1914,\ldots,1918,\ 1939,\ldots,1945$$

during wartime, leading to the scatterplot shown in Figure 17.1.[1]

It is clear that there are two separate models and that the coefficients do not remain constant: the slope shifts down from 0.8, its peacetime value, to 0.5 during the war years, while the intercept also declines, from

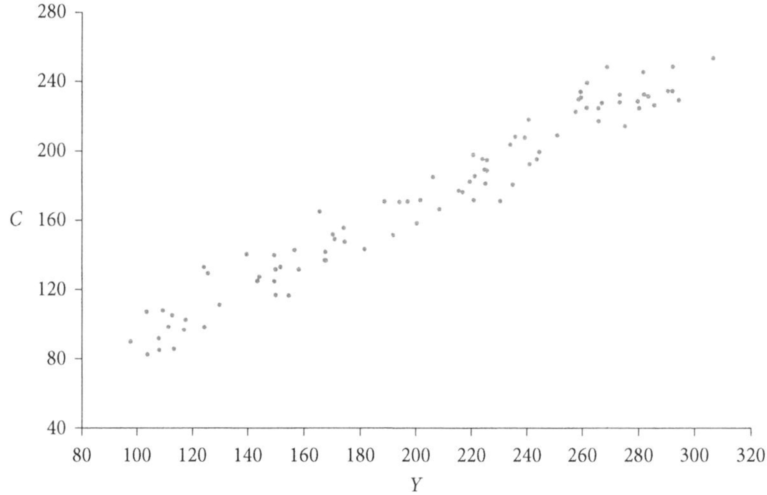

FIGURE 17.1 *'War & Peace' consumption function scatterplot*

10 to 6. Note that the error variance is assumed to remain constant across periods and we do not distinguish between the two world wars, both being assumed to have identical effects. This latter assumption could easily be relaxed in a more general example.

Suppose we ignore the distinction between peacetime and wartime, and estimate a *single* regression. Doing this yields

$$\underset{}{\hat{C}_t = -\underset{(6.74)}{6.93} + \underset{(0.03)}{0.85}\, Y_t}, \quad \hat{\sigma} = 18.9, \quad dw = 0.85$$

The slope coefficient is 'weighted' towards the peacetime MPC of 0.8, as these observations numerically dominate (88 out of the 100 years); but $\hat{\sigma}$ is far too high (it should be around 10) and dw is significant (it should, of course, be around 2 as the errors are $IN(0,100)$ in both periods). The fitted line is superimposed on the scatterplot, shown again in Figure 17.2(a), and also plotted are the residuals (Figure 17.2(b)): wartime consumption is considerably overpredicted, leading to runs of negative residuals and hence a very low dw.

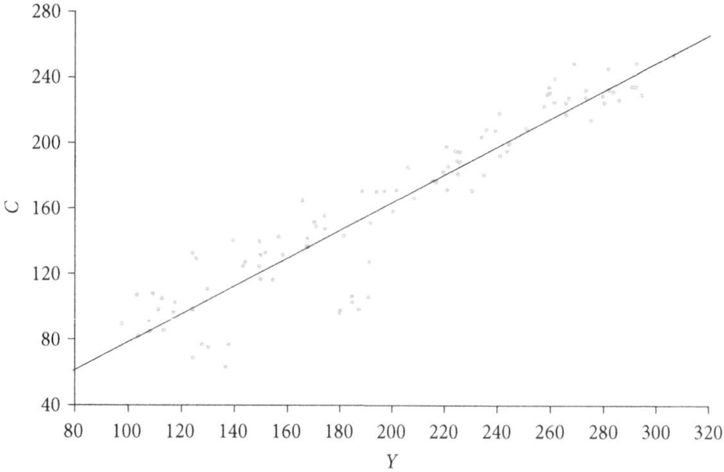

FIGURE 17.2(a) *Consumption function and residuals: single consumption function*

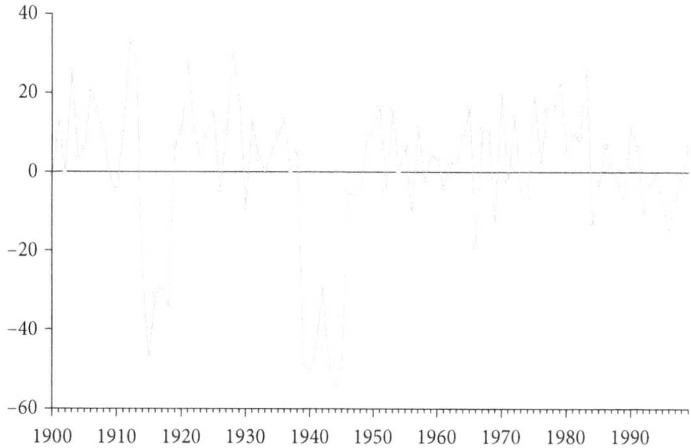

FIGURE 17.2(b) *Consumption function and residuals: residuals*

17.2 A test for structural stability: the Chow test

Suppose that we now have a more general problem. We have two independent sets of data with sample sizes N_1 and N_2, respectively, on the same set of variables, $Y, X_1, \ldots, X_k$ for which we have two models. For the first data set we have

$$Y = \beta_{10} + \beta_{11} X_1 + \beta_{12} X_2 + \ldots + \beta_{1k} X_k + u \qquad (17.1)$$

while for the second we have

$$Y = \beta_{20} + \beta_{21} X_1 + \beta_{22} X_2 + \ldots + \beta_{2k} X_k + u \qquad (17.2)$$

We wish to test the 'stability hypothesis'

$$H_0 : \beta_{10} = \beta_{20},\ \beta_{11} = \beta_{21},\ \beta_{12} = \beta_{22},\ \ldots,\ \beta_{1k} = \beta_{2k}$$

because, if it was true, we could estimate a *single* equation for the two data sets. The hypothesis can be tested using a *Chow test*, which is derived in the following way:[2]

Let RSS_1 be the residual sum of squares from estimating (17.1)

Let RSS_2 be the residual sum of squares from estimating (17.2)

Note that $RSS_1 + RSS_2 = RSS_U$ in the terminology of §**13.3**, since no restrictions are placed on the model by estimating separately over the two data sets.

Let RSS_R be the residual sum of squares from estimating a single regression over all $N_1 + N_2$ observations.

Since the null hypothesis imposes $r = k+1$ restrictions, and there are

$$(N_1 - k - 1) + (N_2 - k - 1) = N_1 + N_2 - 2k - 2$$

degrees of freedom *after* estimation of (17.1) and (17.2), the usual F-statistic from §**13.3** is

$$F = \frac{(RSS_R - RSS_1 - RSS_2)/(k+1)}{(RSS_1 + RSS_2)/(N_1 + N_2 - 2k - 2)} \sim F(k+1, N_1 + N_2 - 2k - 2)$$

In the above consumption function example, we have $k=1$, $N_1 = 88$ peacetime observations and $N_2 = 12$ wartime observations. We also have the following residual sums of squares

$$\left.\begin{aligned} RSS_1 &= 9323.0 \\ RSS_2 &= 689.8 \end{aligned}\right\} \quad RSS_U = 10012.8$$

$$RSS_R = 34906.9$$

and thus

$$F = \frac{34906.9 - 10012.8}{10012.8} \cdot \frac{96}{2} = 119.3 \sim F(2,96)$$

which clearly rejects H_0: $\beta_{10} = \beta_{20}$, $\beta_{11} = \beta_{21}$.

For the Chow test to be applicable, the error variance must be constant across samples, that is, the error needs to be homoskedastic. Hence a Goldfeld–Quandt test (recall §15.4), for example, should be carried out before calculating the Chow test.

A second, and related, test asks whether the model fitted to the first sample can successfully predict the Y observations in the second sample. If it can, then the model has remained unchanged; if it cannot then the coefficients are probably different. This test, known as *Chow's second test*, or the *predictive failure* (PF) test, has the form

$$F_{PF} = \frac{\left(RSS_R - RSS_1\right)/N_2}{RSS_1/\left(N_1 - k - 1\right)} \sim F\left(N_2, N_1 - k - 1\right)$$

For the consumption function,

$$\begin{aligned} F_{PF} &= \frac{(34906.9 - 9323.0)/12}{9323.0/86} \sim F(12,86) \\ &= 19.87 > F_{0.01}(12,86) = 2.39 \end{aligned}$$

and thus, not surprisingly, the model fitted for the peacetime years is unable to predict accurately the values taken by consumption in wartime.

These tests are often used to investigate whether a model has altered after a certain date. For example, we may ask whether our consumption function was different before 1939 than after, so that we have a 'break' at this year, with sub-samples 1900,…,1938 and 1939,…,1999. The tests can be constructed using the above procedures. Note that the PF test can be used when the Chow test cannot, that is, when the 'second' sample has too few observations for a regression to be computed (when $N_2 < k+1$). Even when N_2 is only just larger than $k+1$, the PF test will be preferable, because the degrees of freedom will still only be small.

For reference, the two estimated consumption functions are, for peacetime,

$$\underset{}{\hat{C}_t} = \underset{(3.97)}{10.59} + \underset{(0.019)}{0.795}\, Y_t, \quad \hat{\sigma} = 10.41, \quad dw = 2.01$$

and, for wartime,

$$\hat{C}_t = \underset{(14.62)}{-9.21} + \underset{(0.088)}{0.613}\, Y_t, \quad \hat{\sigma} = 8.31, \quad dw = 2.20$$

The peacetime regression almost reproduces the PRF, but the wartime regression imprecisely estimates the intercept and underestimates σ: this is due to the very small sample size. The two functions are shown superimposed on the scatterplot in Figure 17.3.

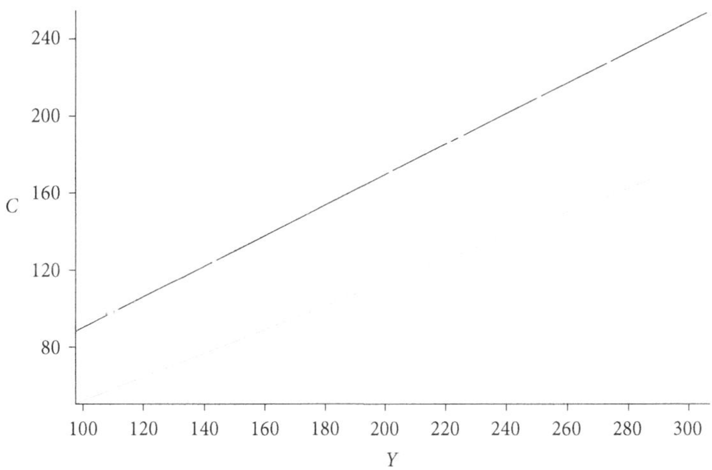

FIGURE 17.3 *Fitted 'War & Peace' consumption functions*

17.3 Using dummy variables to model changing parameters

Continuing with the consumption function example, suppose we define the *dummy variable*

$$D_t = \begin{cases} 0 & \text{if } t \text{ is a peacetime year} \\ 1 & \text{if } t \text{ is a wartime year} \end{cases}$$

Our consumption function is, in general,

$$C_t = \beta_{10} + \beta_{11} Y_t + u_t \qquad \text{peacetime}$$

$$C_t = \beta_{20} + \beta_{21} Y_t + u_t \qquad \text{wartime}$$

Using the dummy variable defined above, these two equations can be written as the *single* equation

$$C_t = \beta_{10} + (\beta_{20} - \beta_{10}) D_t + \beta_{11} Y_t + (\beta_{21} - \beta_{11}) D_t Y_t + u_t$$

because, when $D_t = 0$, the peacetime function is obtained, whereas, if $D_t = 1$,

$$C_t = \beta_{10} + \beta_{20} - \beta_{10} + \beta_{11} Y_t + (\beta_{21} - \beta_{11}) Y_t + u_t = \beta_{20} + \beta_{21} Y_t + u_t,$$

and the wartime function reappears. On defining

$$Z_t = D_t Y_t = \begin{cases} 0 & \text{if } t \text{ is a peacetime year} \\ Y_t & \text{if } t \text{ is a wartime year} \end{cases}$$

and

$$\gamma_0 = \beta_{20} - \beta_{10}, \qquad \gamma_1 = \beta_{21} - \beta_{11},$$

we can now estimate

$$C_t = \beta_{10} + \gamma_0 D_t + \beta_{11} Y_t + \gamma_1 Z_t + u_t$$

If the 'structural stability' hypothesis H_0: $\beta_{10} = \beta_{20}$, $\beta_{11} = \beta_{21}$ holds, then $\gamma_0 = \gamma_1 = 0$, which can, of course, be tested by a standard F statistic: this test is, in fact, exactly the Chow F-test introduced above. We may also test for *just* the intercept shifting ($\gamma_1 = 0$: a *parallel* shift of the function to a new intercept $\beta_{10} + \gamma_0 = \beta_{20}$), or *just* the slope shifting ($\gamma_0 = 0$: the function *pivots* about the intercept β_{10} to the new slope $\beta_{11} + \gamma_1 = \beta_{21}$).

Estimating the 'dummy variable' regression obtains

$$\hat{C}_t = \underset{(3.89)}{10.59} - \underset{(18.40)}{19.80}\ D_t + \underset{(0.018)}{0.795}\ Y_t - \underset{(0.110)}{0.181}\ Z_t$$

$$\hat{\sigma} = 10.21 \qquad dw = 1.99 \qquad RSS = 10012.8$$

which yields the peacetime model when $D_t = Z_t = 0$, and the wartime model when $D_t = 1$ and $Y_t = Z_t$.

Note that in the Chow test the RSS is the same as the RSS_U, the estimate of σ is close to its true value of 10 and the dw is close to its expected value of 2. A t-test of $\gamma_0 = 0$ is insignificant, however, so that we cannot reject the hypothesis that the function pivots about a single intercept. This is consistent with the discussion of the two estimated regressions above and the associated scatterplot.

17.4 Parameter stability: recursive regressions

We have already discussed the two Chow tests for parameter stability. We may also investigate the issue of parameter stability in a rather less formal, but perhaps more informative, manner.

Certain relationships holding in the 'algebra' of OLS may be used to re-estimate coefficients rapidly as the sample size is altered in various ways. One important case is when the sample period is sequentially incremented by one observation at a time, thus leading to the sequence of *recursive residuals* and *coefficients*. More formally, suppose we have the regression

$$Y_t = \beta_0 + \beta_1 X_{1t} + \ldots + \beta_k X_{kt} + u_t, \qquad t = 1,2,\ldots,T$$

As there are k regressors, the minimum number of observations that are required for estimation are $k+2$ (anything less and there will be no degrees of freedom). Setting $r = k+2, k+3,\ldots,T$ the recursive coefficients are denoted as

$$\hat{\beta}_0^{(r)}, \hat{\beta}_1^{(r)},\ldots,\hat{\beta}_k^{(r)}, \quad r = k+2, k+3,\ldots,T$$

these being obtained by estimating the regression initially over the sample $1,2,\ldots,k+2$, then over the sample $1,2,\ldots,k+3$, and so on, so that $\hat{\beta}_i^{(T)}$ is simply the OLS estimator $\hat{\beta}_i$ obtained from estimating over the complete sample $1,2,\ldots,T$. The recursive residuals are defined as

$$\upsilon_r = \left(Y_r - \hat{\beta}_0^{(r-1)} - \hat{\beta}_1^{(r-1)} X_{1r} - \ldots - \hat{\beta}_k^{(r-1)} X_{kr}\right) \Big/ d_r, \quad r = k+3, k+4,\ldots,T$$

that is, they are the residuals obtained by estimating the model over the first r–1 observations and then using the resulting recursive coefficients to predict the next value Y_r (the residuals are divided by the 'scaling factor' d_r, whose formula need not concern us here, to ensure that they have a common variance of σ^2). The recursive residuals can be used to construct two tests which are useful for detecting parameter instability.

The *Cumulative Sum* (CUSUM) statistic is defined as

$$W_r = \sum_{j=k+3}^{r} \upsilon_j \Big/ \hat{\sigma}, \qquad r = k+3, k+4,\ldots,T$$

where $\hat{\sigma}$ is the OLS estimator of σ. The test uses a graphical technique and involves plotting W_r and a pair of straight lines for values of $r = k+3, k+4,\ldots,T$. The straight lines are 5% significance levels, so that if the plot of W_r cuts either of the lines then there is evidence of parameter instability at, or around, the value of r at which the 'break' takes place.

The *Cumulative Sum of Squares* (CUSUM of SQUARES) statistic is defined as

$$WW_r = \sum_{j=k+3}^{r} v_j^2 \bigg/ \sum_{j=k+3}^{T} v_j^2 \qquad r = k+3, k+4, \ldots, T$$

The test again involves plotting WW_r and a pair of 5% significance level lines, and is interpreted in a similar way to the CUSUM test.

Figure 17.4 plots the recursive residuals from the 'War & Peace' consumption function, and Figure 17.5 shows plots of the CUSUM and CUSUM of SQUARES statistics with associated 5% significance level lines.[3]

The recursive residuals are shown accompanied by ±2 standard error bands, and we can see that the v_j penetrate the bands during the first and second world wars, as we should expect. The CUSUM statistic fails to pick up either of the wartime breaks, a consequence of the potential lack of power of this statistic. The CUSUM of SQUARES statistic, however, clearly picks up the break in the second world war, but also fails to identify the break in 1914. Recursive estimates of the intercept and slope are also plotted, in Figure 17.6, and reveal instability in the expected ways.

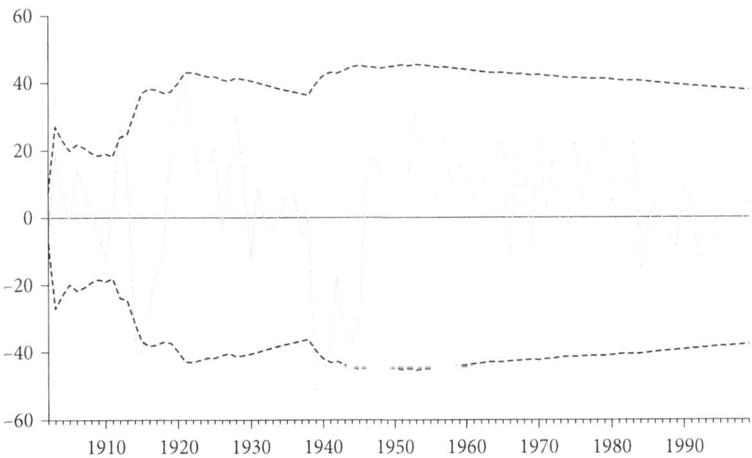

FIGURE 17.4 *Recursive residuals from the consumption function*

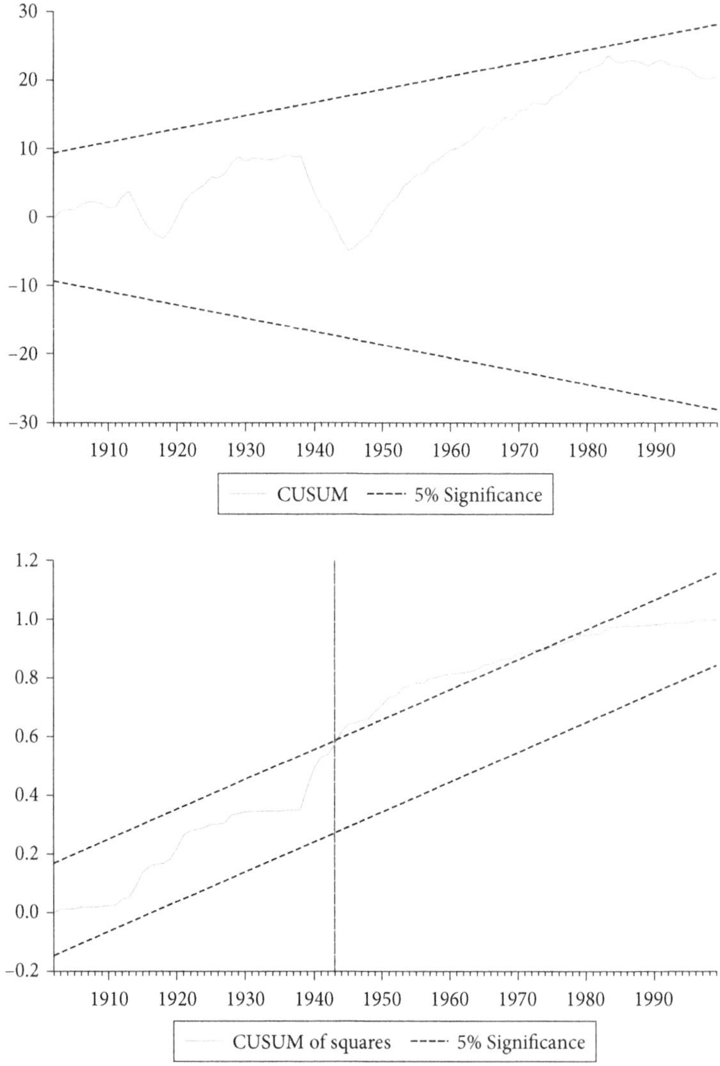

FIGURE 17.5 *CUSUM statistics for the consumption function*

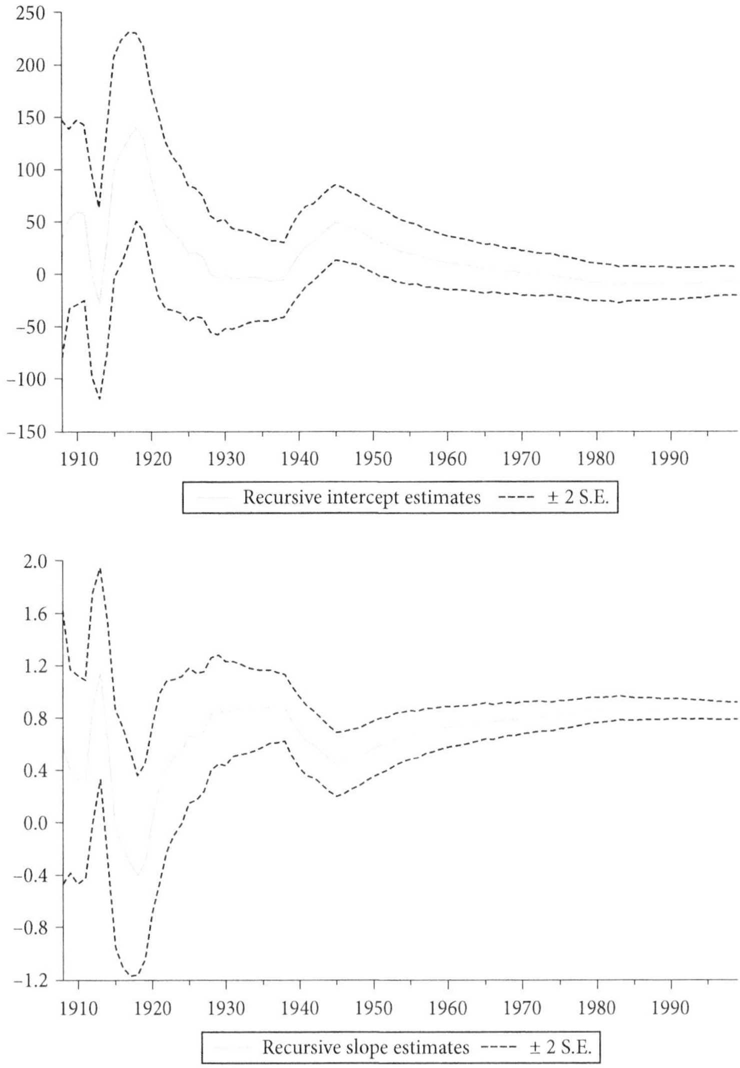

FIGURE 17.6 *Recursive estimates for the consumption function*

17.5 Interest rates and inflation in the UK revisited

Figure 2.6 presented a scatterplot of UK long interest rates and inflation from 1751 to 2011, showing what appeared to be a 'structural shift' in the relationship between the two variables for the period 1965 to 1997: during these years the variables look to be strongly positively correlated, while outside of this period there appears to be little relationship between them at all. We are now in a position to examine this relationship more formally. We begin by reporting regressions between the long interest rate, R_t, and inflation, π_t, and the lagged interest rate, R_{t-1}, for a variety of sample periods:

1751–2011

$$\hat{R}_t = \underset{(0.083)}{0.187} + \underset{(0.006)}{0.010}\,\pi_t + \underset{(0.017)}{0.955}\,R_{t-1} \qquad R^2 = 0.933 \qquad \hat{\sigma} = 0.599$$

$$F(1,257) = 0.51[0.48] \qquad F(2,256) = 0.37[0.69]$$

1751–1964

$$\hat{R}_t = \underset{(0.090)}{0.178} + \underset{(0.003)}{0.004}\,\pi_t + \underset{(0.024)}{0.954}\,R_{t-1} \qquad R^2 = 0.884 \qquad \hat{\sigma} = 0.284$$

$$F(1,210) = 1.79[0.18] \qquad F(2,209) = 3.13[0.05]$$

1965–1997

$$\hat{R}_t = \underset{(1.113)}{4.411} + \underset{(0.066)}{0.243}\,\pi_t + \underset{(0.148)}{0.359}\,R_{t-1} \qquad R^2 = 0.749 \qquad \hat{\sigma} = 1.226$$

$$F(1,29) = 1.92[0.17] \qquad F(2,28) = 3.96[0.06]$$

1998–2011

$$\hat{R}_t = \underset{(0.809)}{4.159} - \underset{(0.058)}{0.062}\,\pi_t + \underset{(0.161)}{0.120}\,R_{t-1} \qquad R^2 = 0.150 \qquad \hat{\sigma} = 0.305$$

$$F(1,10) = 0.28[0.61] \qquad F(2,9) = 0.20[0.82]$$

1751–1964, 1998–2011

$$\hat{R}_t = \underset{(0.092)}{0.288} + \underset{(0.003)}{0.004}\, \pi_t + \underset{(0.024)}{0.922}\, R_{t-1} \qquad R^2 = 0.869 \qquad \hat{\sigma} = 0.303$$

$$F(1{,}224) = 0.88[0.35] \qquad F(2{,}223) = 1.04[0.35]$$

The F-tests are LM tests for first- and second-order residual autocorrelation (§14.8), and show little evidence of dynamic misspecification in the regressions. The estimated coefficients confirm that it is only between 1965 and 1997 that interest rates and inflation are significantly positively related (the t-statistic on π_t is 3.70 for this regression, whereas in the other regressions it is 1.70, 1.50, −1.06 and 1.37, respectively). A Chow test for a break at 1965 produces the highly significant statistic $F(3{,}255) = 9.89$, and the PF test yields the equally significant statistic $F(47{,}211) = 19.99$. If the sample period is restricted to 1965–2011, then the Chow test finds a significant break at 1998 ($F(3{,}41) = 5.90$).

Figures 17.7 and 17.8 show the recursive residuals and the CUSUM of SQUARES plots (the CUSUM plot shows no significance, again suggesting that it often lacks power). It is clear that the years from 1965 to 1997

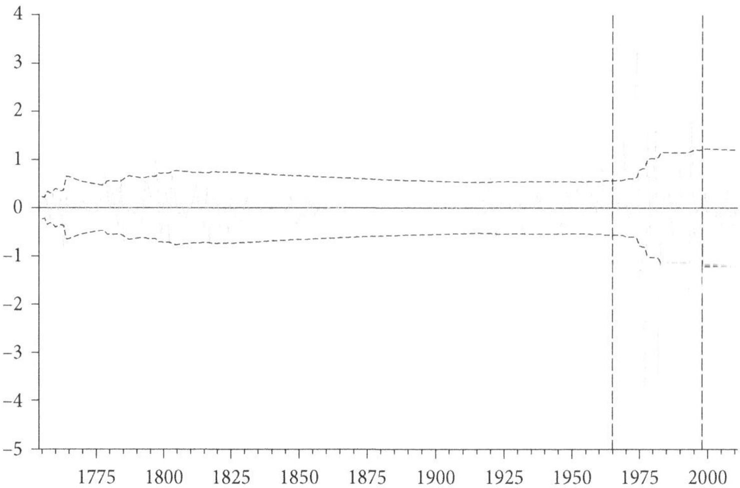

FIGURE 17.7 *Recursive residuals from the interest rate–inflation regression*

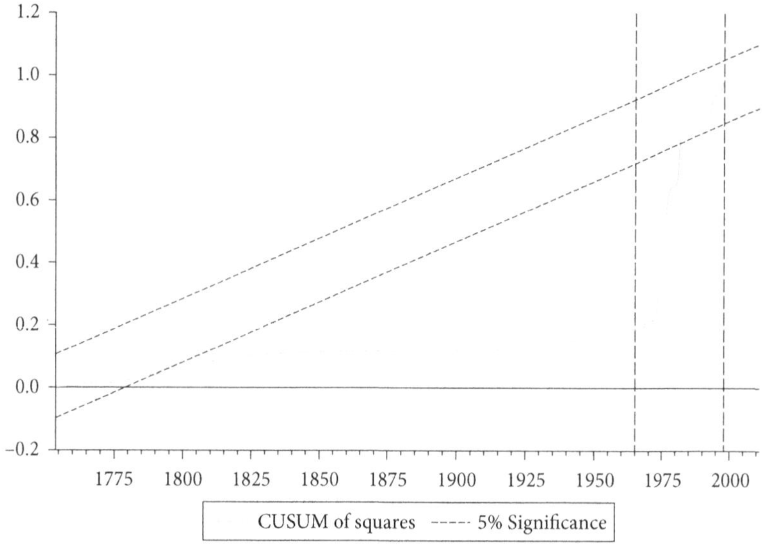

FIGURE 17.8 *CUSUM of SQUARES statistic from the interest rate–inflation regression*

were a period of great instability in the relationship between interest rates and inflation.

From the link between *t*-ratios and partial correlations given in §13.3, we calculate $r_{R\pi.R_{-1}}$ to be 0.54 for 1965–1997, 0.10 for 1751–1964 and −0.27 for 1998–2011, the latter two correlations being insignificant. In fact, on deleting insignificant coefficients in the various regressions, we obtain

1751–1964

$$\hat{R}_t = \underset{(0.090)}{0.177} + \underset{(0.024)}{0.956}\ R_{t-1} \qquad \hat{\sigma} = 0.285$$

1998–2011

$$\hat{R}_t = \underset{(0.081)}{4.548} \qquad \hat{\sigma} = 0.304$$

Since Bank of England independence, therefore, there has been no relationship between interest rates and inflation, with the long interest rate fluctuating randomly around a mean of approximately 4½%. No relationship between the variables was also the norm before 1965: however, in this period there was a high degree of persistence in the interest rate, with the lagged interest rate having a coefficient only a little below unity. Indeed, imposing the value of unity on this coefficient yields[4]

$$\Delta\hat{R}_t = \underset{(0.020)}{0.014} \qquad \hat{\sigma} = 0.285$$

and the insignificance of the intercept in this regression implies that up until 1965 the interest rate followed a drift-free random walk (recall §**6.6**). For the 'aberrant' 1965–1997 period, the relationship is approximately $R_t = 4.5 + 0.25\pi_t + 0.4R_{t-1}$. We are thus led to the view that the process generating the interest rate is *regime switching*, altering because of economic and institutional changes during the last half of the 20th century.[5]

Notes

1 The income data is assumed to be generated by $Y_t = 100 + 2t + v_t$, $v_t \sim IN(0,25)$.

2 Gregory C. Chow, 'Tests of equality between sets of coefficients in two linear regressions', *Econometrica* 28 (1960), 591–605.

3 These statistics were proposed by R.L. Brown, James Durbin and D.M. Evans, 'Techniques for testing for the constancy of regression relationships over time', *Journal of the Royal Statistical Society, Series B* 37 (1975), 141–192, where expressions for the standard errors associated with the two statistics may be found.

4 Testing the hypothesis that the coefficient on R_{t-1} is unity through a *t*-test turns out to be invalid, because under this null hypothesis the *t* statistic does not follow a *t*-distribution. This problem of testing for a *unit root* is discussed in §**18.7**.

5 In fact, there is evidence that the interest rate-generating process altered even more frequently over this period, in the earlier years being dependent upon whether Britain was on the gold standard or not. Much more detailed statistical and economic analysis of the historical interaction between inflation and interest rates is provided by Terence C. Mills, 'Exploring historical economic relationships: two and a half centuries of British interest rates and inflation', *Cliometrica* 2 (2008), 213–228, and Terence C. Mills and Geoffrey E. Wood, 'Two and a half centuries of British interest rates, monetary regimes and inflation', in Nicholas Crafts, Terence C. Mills and Geoffrey E. Wood (editors), *Explorations in Financial and Monetary History: Essays in Honour of Forrest H. Capie* (London, Routledge, 2011), pp. 158–177.

18
Basic Time Series Models

Abstract: *A popular model to describe an economic time series is that of an autoregression, in which the current value is expressed as a function of past values. This is a simple class of time series model and methods of determining the order of an autoregression are considered. Moving average and mixed models may also be fitted, and methods for building such models are developed by way of several empirical examples. An important requirement for time series modelling is that of stationarity, and the use of differencing and of formal testing procedures for inducing such a property are both considered. It is also important to distinguish between various types of non-stationarity, and the trend stationarity versus difference stationarity distinction is developed in some detail.*

18.1 Autoregressions

In §6.6 the lagged dependent variable regression was introduced, taking the form

$$X_t = a + bX_{t-1} + u_t \tag{18.1}$$

This is, in fact, an example of a *first-order autoregression*, often referred to as an AR(1) process.[1] A natural extension is the general p-th order autoregression (AR(p))

$$X_t = \theta + \phi_1 X_{t-1} + \ldots + \phi_p X_{t-p} + u_t \tag{18.2}$$

in which X_t is regressed on p lagged values of itself: hence the term autoregression. An equivalent 'difference from mean' form is

$$(X_t - \mu) = \phi_1 (X_{t-1} - \mu) + \ldots + \phi_p \left(X_{t-p} - \mu\right) + u_t \tag{18.3}$$

where the relationship between the intercept θ in (18.2) and the mean of X_t, μ, in (18.3) is given by $\theta = \mu(1-\phi_1-\ldots-\phi_p)$. The errors u_t are assumed to satisfy the classical regression assumptions of §12.2 and, within this explicitly time series context, u_t is often referred to as a *white noise innovation*, denoted $u_t \sim WN(0,\sigma^2)$.[2]

18.2 Stationarity

It was pointed out in §6.6 that if $b=1$ in (18.1) then X_t follows a random walk with drift. This is an example of a *non-stationary* process and it is easy to show that if X_0 is the initial value of the series, then $E(X_t) = X_0 + ta$ and $V(X_t) = t\sigma^2$, so that both the mean and variance of X_t are time-dependent.[3] This makes statistical analysis rather problematic, and it is therefore typically assumed that X_t is *stationary*, which requires in the AR(1) case that $|b| < 1$ and, generally, that $\phi_1 + \ldots + \phi_p < 1$. Stationarity rules out the presence of trends in the data (cf. the trends in consumption and income examined in §6.6), and so if such series are to be analysed within an autoregressive framework, they must first be transformed to stationarity. Consequently, if generated by the AR(1) process (18.1), the stationary series X_t will have a constant mean of $\mu = a/(1-b)$ and a constant variance of $\sigma^2/(1-b^2)$.[4] The operation of *differencing*, that is, of defining $\Delta X_t = X_t - X_{t-1}$, has been found to be a useful way of inducing stationarity in many time series, and when combined with the taking of logarithms essentially transforms a non-stationary, trending series to a stationary series of growth rates: recall using such a transformation in §3.2 to transform the non-linearly trending RPI to stationary inflation.

18.3 Determining the order of an autoregression

Important and useful statistics for modelling time series are the *sample autocorrelations* and *sample partial autocorrelations*. The *lag-k sample autocorrelation* is defined for the sample $X_1,\ldots,X_T$ as

$$r_k = \frac{\sum_{t=k+1}^{T}\left(X_t - \bar{X}\right)\left(X_{t-k} - \bar{X}\right)}{\sum_{t=1}^{T}\left(X_t - \bar{X}\right)^2}$$

and may be interpreted as an estimator of the corresponding population (or theoretical) autocorrelation

$$\rho_k = \frac{Cov\left(X_t, X_{t-k}\right)}{V\left(X_t\right)}$$

which uses the stationarity implication that $V(X_t) = V(X_{t-k})$ and illustrates the fact that under stationarity the autocorrelations and autocovariances depend solely on the 'time-lag' k and not on time t.

The *lag-k sample partial autocorrelation* is defined as the regression coefficient $\hat{\phi}_{kk}$ in the estimated k-th order autoregression

$$\hat{X}_t = \hat{\theta} + \hat{\phi}_{11} X_{t-1} + \ldots + \hat{\phi}_{kk} X_{t-k}$$

and may be interpreted as an estimate of the coefficient on X_{t-k} in an AR(k) process, ϕ_{kk}, which is the theoretical partial autocorrelation. Listings of the sample autocorrelations and sample partial autocorrelations with respect to k are known as the *sample autocorrelation function* (SACF) and *sample partial autocorrelation function* (SPACF) respectively, while analogous listings of their theoretical counterparts are termed the ACF and PACF.

The partial autocorrelations may be used to select the order of an autoregression, and this makes models such as (18.3) operational, since on selecting a value for p, the autoregression may be estimated by OLS. If X_t was really generated by an AR(k) process, then fitting autoregressions of higher order would lead to the $\hat{\phi}_{k+1,k+1},\ldots$ all being insignificantly different from zero. Since the standard error of all sample partial autocorrelations of lag greater than k is approximately $1/\sqrt{T}$, then as long as T is reasonably large, standard t-tests may be constructed to determine the significance of $\hat{\phi}_{kk}$. This procedure may be performed iteratively, with the order of the autoregression being determined as that lag after which all higher order partial autocorrelations are insignificantly different from zero.[5]

An alternative approach to selecting p is to choose a maximum autoregressive order – K, say – and to fit all autoregressions up to that order. If the estimated error variance of the fitted k-th order autoregression is denoted $\hat{\sigma}_k^2$, then the order of the autoregression may be selected as that value which minimises the *information criterion*

$$IC_k = \ln\hat{\sigma}_k^2 + f(k,T)T^{-1}, \qquad k = 1,\ldots,K$$

Two popular choices for the 'penalty function' are $f(k,T) = 2k$, which defines Akaike's Information criterion (AIC), and $f(k,T) = k\ln T$, which defines Schwarz's Bayesian Criterion (BIC).[6]

Determining the order of an autoregression for inflation

In **§6.6** an AR(1) process was fitted to annual inflation from 1948. The SPACF, AIC and BIC values for $k \leq 10$ are shown in Table 18.1.

Since $T = 63$ the standard error of each $\hat{\phi}_{kk}$ is 0.126 so that only $\hat{\phi}_{11}$ is significantly different from zero. The choice of an AR(1) process is

TABLE 18.1 *Autoregressive order determination statistics for inflation*

k	$\hat{\phi}_{kk}$	AIC_k	BIC_k
1	0.775	5.056	5.125
2	−0.087	5.062	5.166
3	0.201	5.067	5.207
4	0.171	5.037	5.213
5	−0.004	5.090	5.304
6	−0.232	5.009	5.260
7	−0.140	5.051	5.340
8	−0.068	5.103	5.432
9	0.032	5.155	5.523
10	−0.014	5.196	5.605

confirmed by the *BIC*, but the *AIC* is minimised at $k=6$. The *AIC* typically selects a higher order than the *BIC*, as the latter's penalty function penalises higher values of k more onerously than the former. The fitted AR(1) and AR(6) models are

$$\underset{}{\hat{\pi}_t} = \underset{(0.60)}{1.23} + \underset{(0.081)}{0.776}\,\pi_{t-1} \qquad \hat{\sigma}_u = 2.984$$

$$\hat{\pi}_t = \underset{(0.67)}{1.14} + \underset{(0.135)}{0.832}\,\pi_{t-1} - \underset{(0.183)}{0.064}\,\pi_{t-2} - \underset{(0.181)}{0.048}\,\pi_{t-3} \qquad \hat{\sigma}_u = 2.797$$

$$+ \underset{(0.179)}{0.118}\,\pi_{t-4} + \underset{(0.178)}{0.260}\,\pi_{t-5} - \underset{(0.125)}{0.319}\,\pi_{t-6}$$

The estimated coefficients on lags 2, 3 and 4 in the latter model are very imprecisely determined, and deleting these regressors leads to the 'restricted AR(6)' model

$$\hat{\pi}_t = \underset{(0.63)}{1.14} + \underset{(0.089)}{0.785}\,\pi_{t-1} + \underset{(0.126)}{0.341}\,\pi_{t-5} - \underset{(0.125)}{0.319}\,\pi_{t-6} \qquad \hat{\sigma}_u = 2.734$$

$$AIC = 4.917 \qquad\qquad BIC = 5.060$$

The *AIC* has been reduced even further, and the model has a *BIC* that is smaller than any other model so far fitted.

18.4 Checking the fit of an autoregression

Analogous to checking the fit of a regression model, as detailed in Chapters 14–17, the residuals from a fitted autoregression should 'mimic' white noise and, in particular, should contain no autocorrelation. An obvious test statistic to use is the LM test of §14.8, but an alternative is to consider the *portmanteau test statistic* based on the *residual autocorrelations*

$$\hat{r}_k = \frac{\sum_{t=k+1}^{T}\hat{u}_t\hat{u}_{t-k}}{\sum_{t=1}^{T}\hat{u}_t^2} \qquad \hat{u}_t = X_t - \hat{\theta} - \hat{\phi}_1 X_{t-1} - \ldots - \hat{\phi}_p X_{t-p}$$

This statistic is defined as

$$Q(m) = T(T+2)\sum_{k=1}^{m}(T-k)^{-1}\hat{r}_k^2 \sim \chi^2(m-p)$$

and tests the null hypothesis that the first m residual autocorrelations are all zero.[7] Individual residual autocorrelations may also be tested by comparing them with their standard error of $1/\sqrt{T}$.

Diagnostic checking of autoregressive models of inflation

The AR(1) model has $\hat{r}_5 = 0.318$ and $Q(5) = 11.66$. Not only are both highly significant, but also they point the way to including lags of up to six in the autoregression. The AR(6) fits, either unrestricted or restricted, offer no evidence of any residual autocorrelation using these tests.

18.5 Moving average and mixed time series models

A distinctive feature of autoregressive processes is that their ACFs tail off with k, following a mixture of geometric declines and damped sine waves. For example, the autocorrelations of the AR(1) process (18.1) are given by $\rho_k = b^k$, which will follow a geometric decline under the stationarity assumption that $|b| < 1$.[8] For an AR(p) process the ACF is given by

$$\rho_k = A_1 g_1^k + A_2 g_2^k + \ldots + A_p g_p^k \tag{18.4}$$

where $g_1^{-1}, g_2^{-1}, \ldots, g_p^{-1}$ are the roots of the *characteristic equation*, the p-th order polynomial

$$z^p - \phi_1 z^{p-1} - \ldots - \phi_p = (1 - g_1 z)\ldots(1 - g_p z) = 0$$

and $A_1, A_2, \ldots, A_p$ are constants that depend upon the g_is, $i = 1, 2, \ldots, p$. Stationarity requires that $|g_i| < 1$, so that if a root is real, the term $A_i g_i^k$ in (18.4) geometrically decays to zero as k increases. If there is a complex pair of roots then these will contribute a term to (18.4) which follows a damped sine wave.

The SACF of an autoregressive process will display analogous features, while the PACF and SPACF will 'cut off' at the order of the process, p. Thus Figure 18.1 shows the SACF and SPACF of inflation along with

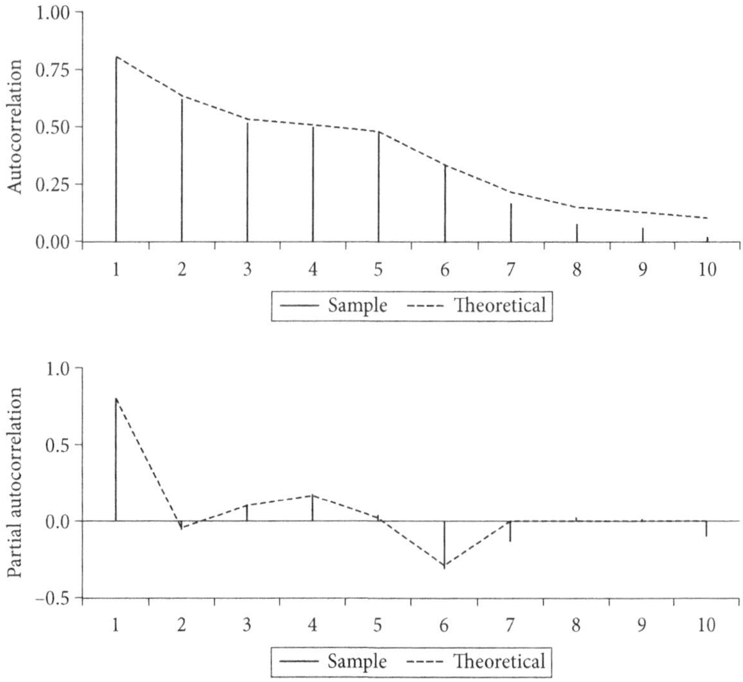

FIGURE 18.1 *The sample ACF and PACF of inflation and the theoretical ACF and PACF from the fitted AR(6) model*

their 'theoretical' counterparts obtained from the fitted AR(6) model, and these clearly exhibit both these features of autoregressive processes.

What, though, if the reverse of this is observed: that is, the SACF appears to cut off but the SPACF tails off? This will happen if X_t follows a *moving average* (MA) process. A first-order moving average (MA(1)) takes the form $X_t = \theta + u_t + \vartheta u_{t-1}$ so that $\rho_1 = \vartheta/(1+\vartheta^2)$ and $\rho_k = 0$ for $k>1$, that is, the ACF cuts off at lag one and the memory of the process is just one period.[9] A moving average process is *always* stationary, but since ρ_1 and ϑ are linked by the quadratic $\rho_1\vartheta^2 - \vartheta + \rho_1 = 0$, which will have solutions

$$\vartheta_1, \vartheta_2 = \frac{-1 \pm \sqrt{1-4\rho_1^2}}{2\rho_1}$$

there will be two solutions for ϑ for a particular value of ρ_1: since $\vartheta_1\vartheta_2 = 1$, these are ϑ_1 and $\vartheta_2 = \vartheta_1^{-1}$. This lack of 'identification' is typically resolved by taking the *invertible* solution $|\vartheta_1| < 1$, which is known as the *invertibility condition*.

The PACF for an MA(1) process can be shown to be dominated by a geometric decline and thus there will be a *duality* existing between AR(1) and MA(1) processes.[10] The ACF of an MA(1) process has a cut-off after lag one, but the ACF of an AR(1) process declines geometrically. Conversely, whereas the PACF of an MA(1) process tails off and is dominated by a geometric decline , the PACF of an AR(1) process has a cut-off after lag 1.

A similar duality holds for the MA(q) process

$$X_t = \theta + u_t + \vartheta_1 u_{t-1} + \ldots + \vartheta_q u_{t-q} \tag{18.5}$$

when compared to an AR(p) process: (18.5) will have an ACF that cuts off after q lags and a PACF that behaves like that of the ACF of an AR(p) process in that it will tail off as a mixture of geometric declines and damped sine waves.

A moving average model for inflation

Figure 18.2 shows the SACF and SPACF of the inflation series shown in Table 2.2, plotted as Figure 2.2 and used in the example in §17.5, along with the 'theoretical' ACF and PACF from the fitted MA(1) process[11]

$$\pi_t = \underset{(0.562)}{2.246} + \hat{u}_t + \underset{(0.054)}{0.476}\ \hat{u}_{t-1} \qquad \hat{\sigma} = 6.158$$

The model fits the low-order sample autocorrelations and partial autocorrelations quite well, and implies that the 'memory' of the inflation process is one year; inflation values a year apart are positively correlated ($r_1 = 0.47$), but values further than a year apart are uncorrelated.

However, it can be seen that some modestly sized higher-order autocorrelations are not so closely fitted, and a more complicated model might therefore be entertained (this is confirmed by a Q(5) statistic of 15.28: in general, if an MA(q) process is fitted then $Q(m) \sim \chi^2(m-q)$).

Two further possibilities exist: either both the SACF and SPACF decline or they both cut off. In the former case a *mixed autoregressive-moving average* (ARMA) process might be considered. The ARMA(1,1) process is

$$X_t = \theta + \phi X_{t-1} + u_t + \vartheta u_{t-1} \tag{18.6}$$

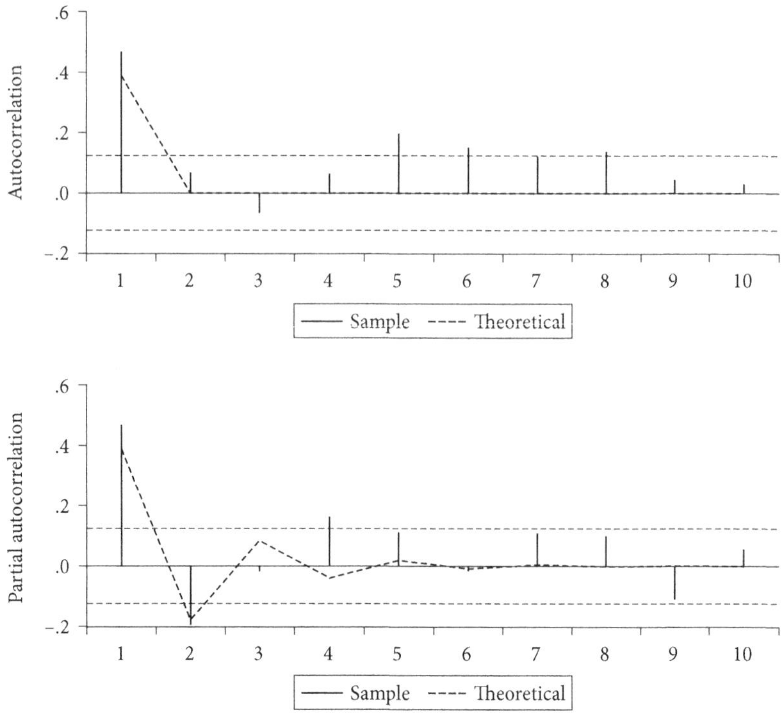

FIGURE 18.2 *The sample ACF and PACF of inflation and the theoretical ACF and PACF from the fitted MA(1) model with two standard error bounds of 0.12*

For stationarity it is required that $|\phi|<1$, while for invertibility $|\vartheta|<1$. The ACF will decline geometrically as $\rho_k = \phi^k$ for $k>1$, with the lag-one autocorrelation being given by[12]

$$\rho_1 = \frac{(1+\phi\vartheta)(\phi+\vartheta)}{(1+\vartheta^2+2\phi\vartheta)}$$

The PACF also has a single initial value, $\phi_{11} = \rho_1$, but then behaves as an MA(1) process, being dominated by a geometric decline. It is rare when modelling economic time series to find that more complicated mixed models than the ARMA(1,1) are needed.

An ARMA model for inflation

An autoregressive term was added to the MA(1) specification for inflation, producing the fitted ARMA(1,1) model

$$\pi_t = \underset{(0.656)}{2.279} + \underset{(0.117)}{0.250}\pi_{t-1} + \hat{u}_t + \underset{(0.116)}{0.298}\hat{u}_{t-1} \qquad \hat{\sigma} = 6.113$$

This offers a modest improvement over the MA(1), but in fact does little to improve the fit of the higher-order autocorrelations.

A potential problem when modelling this inflation series is that the presence of large outliers (recall the histogram shown in Figure 2.3 and the boxplot of Figure 2.5) leads to non-normality of the residuals, with the Jarque–Bera statistic of **§16.6** being $JB = 127.0$. Following the suggestion of that section, two dummy variables were defined, $D_{1800,t}$ and $D_{1802,t}$, and included in the ARMA(1,1) specification, leading to

$$\pi_t = \underset{(0.650)}{2.289} + \underset{(4.95)}{24.80}\, D_{1800,t} - \underset{(4.94)}{26.02}\, D_{1802,t} + \underset{(0.112)}{0.356}\, \pi_{t-1} + \hat{u}_t + \underset{(0.118)}{0.199}\, \hat{u}_{t-1}$$

with $\hat{\sigma} = 5.636$ and $JB = 32.9$, which, although still significant, is much reduced.[13]

The situation when both the SACF and SPACF cut off, typically at low lags, is very common when analysing economic time series. Figure 18.3 shows the SACF and SPACF of the output growth series shown in the scatterplot of Figure 5.3, along with the theoretical ACFs and PACFs from the following fitted AR(2) and MA(1) processes:

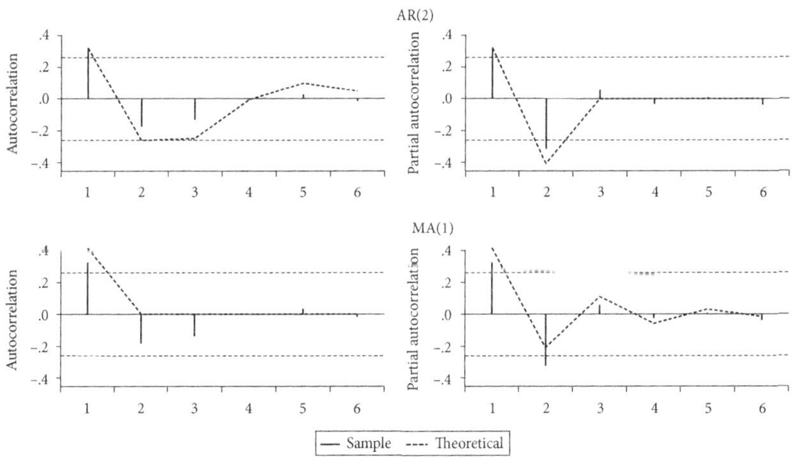

FIGURE 18.3 *ACFs and PACFs from AR(2) and MA(1) fits to output growth with two standard error bounds of 0.26*

$$X_t = \underset{(0.440)}{2.325} + \underset{(0.125)}{0.453}\ X_{t-1} - \underset{(0.139)}{0.407}\ X_{t-2} + \hat{u}_t \qquad \hat{\sigma} = 1.806$$

$$X_t = \underset{(0.350)}{2.429} + \hat{u}_t + \underset{(0.112)}{0.535}\ \hat{u}_{t-1} \qquad \hat{\sigma} = 1.803$$

Only r_1, $\hat{\phi}_{11}$ and $\hat{\phi}_{22}$ are significant, and the fits of the two models are very similar. The reason for this similarity is that an MA(1) process can be written as

$$\begin{aligned}
X_t &= \theta + u_t + \vartheta u_{t-1} = \theta + u_t + \vartheta\left(X_{t-1} - \vartheta u_{t-2} - \theta\right) \\
&= \theta(1-\vartheta) + \vartheta X_{t-1} + u_t - \vartheta^2 u_{t-2} = \theta\,(1-\vartheta) + \vartheta X_{t-1} \\
&\quad + u_t - \vartheta^2\left(X_{t-2} - \vartheta u_{t-3} - \theta\right) \\
&= \theta\left(1 - \vartheta + \vartheta^2\right) + \vartheta X_{t-1} - \vartheta^2 X_{t-2} + u_t + \vartheta^3 u_{t-3} \\
&\ \ \vdots \\
&= \theta\left(1 + (-\vartheta) + (-\vartheta)^2 + \ldots + (-\vartheta)^k\right) - (-\vartheta)X_{t-1} - (-\vartheta)^2 X_{t-2} \\
&\quad - \ldots - (-\vartheta)^k X_{t-k} + u_t - (-\vartheta)^k u_{t-k}
\end{aligned}$$

Using the invertibility condition $|\vartheta| < 1$ then, as $k \to \infty$,

$$X_t = \frac{\theta}{1+\vartheta} + \vartheta X_{t-1} - \vartheta^2 X_{t-2} + \ldots - (-\vartheta)^k X_{t-k} - \ldots + u_t$$

In other words, an invertible MA(1) process can be approximated by a stationary autoregression of infinite order. With $\vartheta = 0.54$, so that $\vartheta^2 = 0.29$ and $\vartheta^3 = 0.16$, the fitted AR(2) process will thus give a reasonable approximation to the fitted MA(1) model.

18.6 Differencing to stationarity and beyond

Figure 18.4 provides the SACF and SPACF of the dollar/sterling exchange rate shown in Figure 3.8, along with the theoretical ACF and PACF obtained from the fitted model implied by the SACF and SPACF, the AR(2) process $X_t = 0.039 + 1.110X_{t-1} - 0.134X_{t-2} + u_t$.

A feature of the SACF is its slow, essentially linear, decline. This may be explained by noting that for an AR(2) process with one root close to the non-stationary boundary and the other root small (the two roots are estimated here to be 0.97 and 0.14), (18.4) can be written as

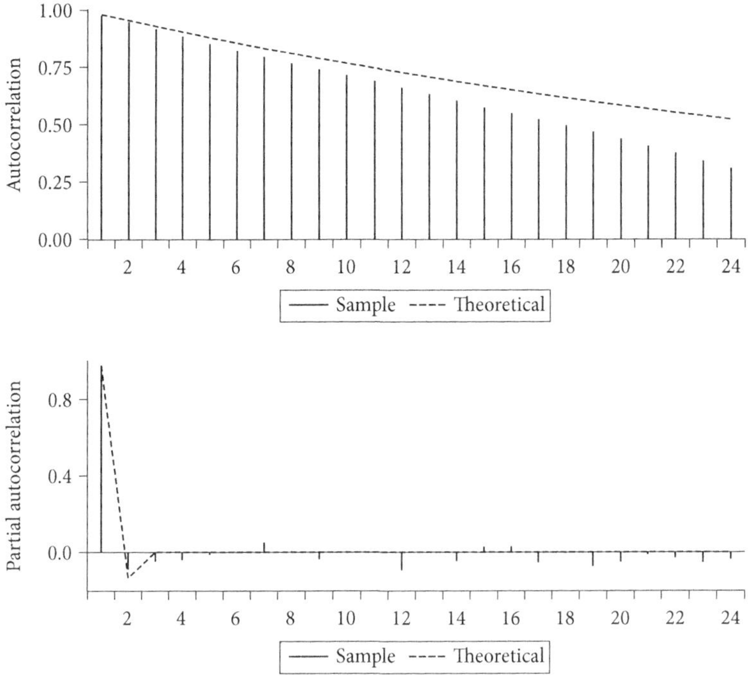

FIGURE 18.4 *The sample ACF and PACF of the exchange rate and the theoretical ACF and PACF from the fitted AR(2) model*

$$\rho_k = A_1 g_1^k + A_2 g_2^k \approx A_1(1-\delta)^k = A_1\left(1-k\delta+k^2\delta^2-k^3\delta^3+\ldots\right) \approx A_1(1-k\delta)$$

for δ small. Since in these circumstances $A_1 \approx 1$, the autocorrelations decline approximately linearly as $\rho_1 = 1-\delta$, $\rho_2 = 1-2\delta = \rho_1 - \delta, \ldots$, $\rho_k = \rho_1 - (k-1)\delta$.[14]

The closeness of the largest root to unity suggests that the exchange rate is borderline non-stationary at best, and would be better treated as being non-stationary, leading to the conclusion that the SACF and SPACF of the differences $\Delta X_t = X_t - X_{t-1}$ should be examined. In fact, this makes sense economically, for the 'difference from mean' form of the AR(2) model is

$$X_t - 1.677 = 1.110\left(X_{t-1} - 1.677\right) - 0.134\left(X_{t-2} - 1.677\right) + u_t$$

This has the implication that the exchange rate always reverts, albeit slowly, to an 'equilibrium' value of 1.677 dollars to the pound, and would thus offer traders a 'one-way bet' whenever the rate got too far away from this value. Such bets do not happen in the foreign exchange market – or indeed, it is argued in the *theory of efficient markets*, in any financial market.

Figure 18.5 shows the ACFs and PACFs from an MA(1) fitted to the changes in the exchange rate: these are almost the same as those obtained from an AR(1) fit, since the models are essentially identical using the analysis of §18.5:

$$\underset{}{\Delta X_t} = -\underset{(0.0026)}{0.0017} + \hat{u}_t + \underset{(0.046)}{0.116}\,\hat{u}_{t-1} \qquad \hat{\sigma} = 0.0511$$

$$\Delta X_t = -\underset{(0.0460)}{0.0016} + \underset{(0.046)}{0.125}\,\Delta X_{t-1} + \hat{u}_t \qquad \hat{\sigma} = 0.0511$$

Note that the intercepts are estimated to be both small and insignificant and hence can be omitted. This, too, makes good economic sense, because the presence of an intercept would imply that exchange rate changes were, on average, non-zero so that over time the exchange rate

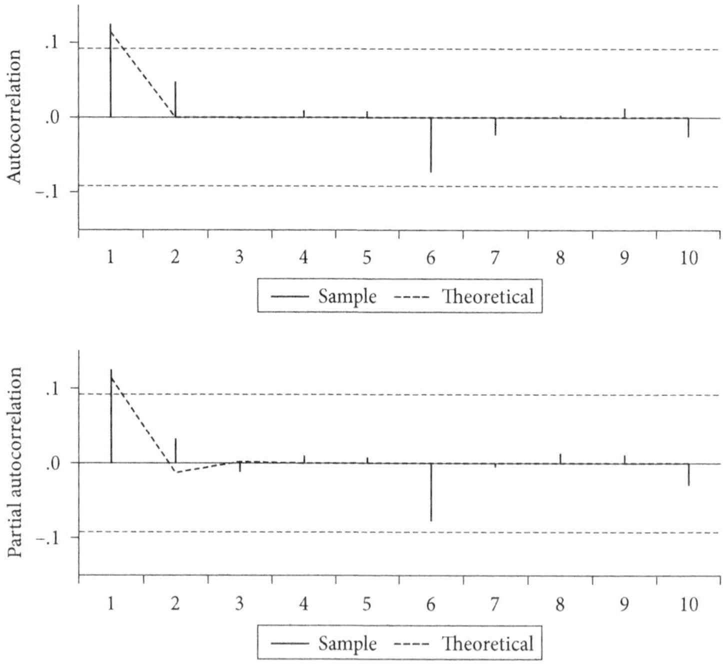

FIGURE 18.5 *ACFs and PACFs from an MA(1) fitted to changes in the exchange rate with two standard error bounds of 0.09*

itself would be continually appreciating or depreciating, again producing a one-way bet that is inconsistent with efficient market theory. Indeed, the two models fitted to the changes in the exchange rate are examples of an *autoregressive-integrated-moving average*, or ARIMA (p,d,q), model, where d denotes the *order of integration*, the number of times the series needs to be differenced to transform it to stationarity. The models here are either ARIMA(1,1,0) or ARIMA(0,1,1).

What happens when an already stationary series is differenced, an occurrence known as *over-differencing*? To take a simple example, suppose that a series is itself white noise, so that $X_t = u_t$. On differencing, this becomes $\Delta X_t = u_t - u_{t-1}$: in other words, ΔX_t follows an MA(1) process that, since $\vartheta = -1$, is not invertible. However, since all moving average processes are stationary, ΔX_t is stationary. Note that the lag-one autocorrelation is $\rho_1 = -0.5$, so that a good indication of over-differencing will be a first-order sample autocorrelation in the region of this value, accompanied by an SPACF that declines slowly from −0.5.

Figure 18.6 shows the consequences of over-differencing the exchange rate.

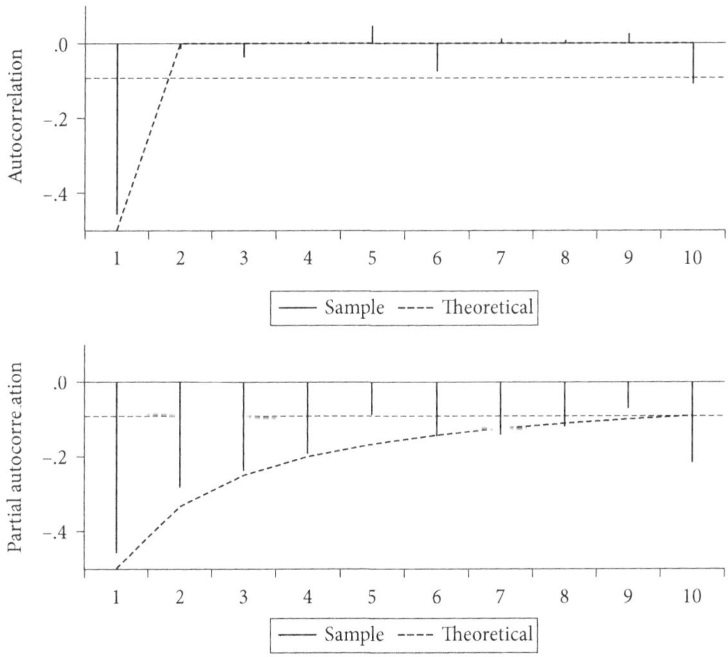

FIGURE 18.6 *ACFs and PACFs of the second differences of the exchange rate*

This figure presents the SACF and PACF of the *second differences* $\Delta^2 X_t = \Delta(\Delta X_t) = X_t - 2X_{t-1} + X_{t-2}$, along with the theoretical ACF and PACF of the fitted ARIMA(0,2,1) model

$$\Delta^2 X_t = \hat{u}_t - \underset{(0.003)}{0.996}\, \hat{u}_{t-1} \qquad \hat{\sigma} = 0.0515$$

With $r_1 = -0.46$, $\hat{\vartheta} = -0.996$ and an approximately geometrically declining SPACF, all the implied consequences of over-differencing are clearly seen. Note also that the estimate of σ has increased on fitting an over-differenced model: this is often an indication of the problem.

18.7 Testing for unit roots

Determining the appropriate order of differencing, or degree of integration, is not always straightforward from just an examination of the SACFs of various differences of a time series. Consider the series shown in Figure 18.7.

This is the 'spread' between the long and short UK interest rates examined in **§16.6**, that is, it is defined as $X_t = R_t - r_t$. From a visual inspection, two possibilities suggest themselves: either the spread is stationary, in

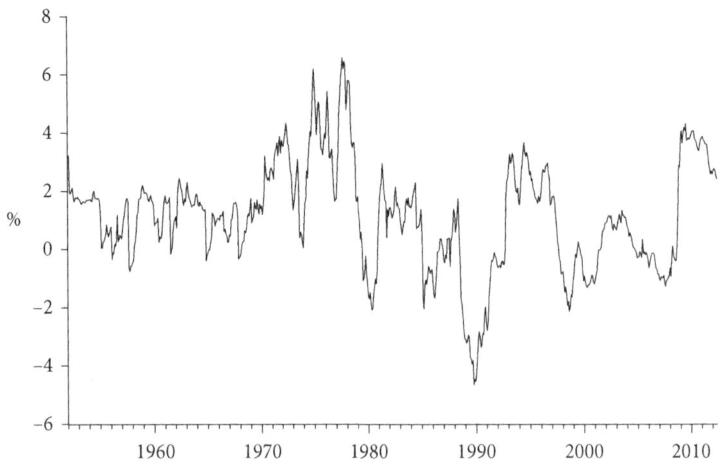

FIGURE 18.7 *Spread between long and short UK interest rates, March 1952–June 2012*

which case it fluctuates rather persistently around a constant mean which has the interpretation of being the equilibrium difference between long and short interest rates; or it is non-stationary and thus wanders randomly, in which case the concept of an equilibrium difference between long and short rates has no meaning.

The SACFs and SPACFs of the levels and first differences of the spread, along with their theoretical counterparts obtained by fitting the following ARIMA(2,0,0) and ARIMA(1,1,0) models to X_t, are shown in Figure 18.8.

$$X_t = \underset{(0.018)}{0.034} + \underset{(0.036)}{1.191}\ X_{t-1} - \underset{(0.036)}{0.223}\ X_{t-2} + \hat{u}_t \qquad \hat{\sigma} = 0.4081$$

$$\Delta X_t = \underset{(0.036)}{0.207}\ \Delta X_{t-1} + \hat{u}_t \qquad \hat{\sigma} = 0.4118$$

The ARIMA(2,0,0) model implies that the spread fluctuates around a mean of $\mu = 0.034/(1-1.191+0.223) = 1.07$ percentage points as an AR(2) process with roots of 0.96 and 0.23. The ARIMA(1,1,0) says that changes in the spread follow an AR(1) process with a root of 0.21.

The essential difference between the two models is that the latter imposes a *unit root*, whereas the former estimates this largest root to

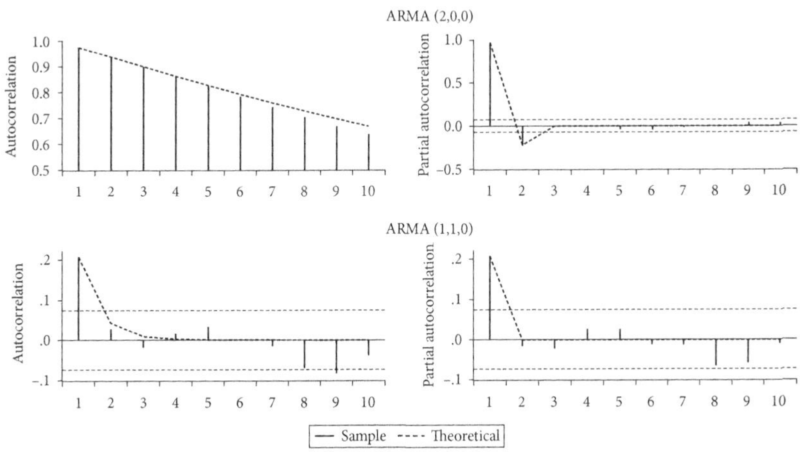

FIGURE 18.8 *ACFs and PACFs from ARIMA(2,0,0) and ARIMA(1,1,0) fits to the spread with two standard error bounds of 0.07*

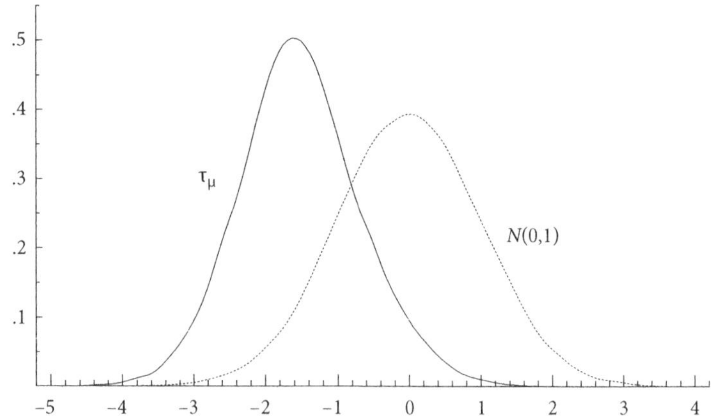

FIGURE 18.9 *The Dickey–Fuller τ_μ distribution compared to the standard normal*

be 0.96. What is thus required to formally distinguish between the two models is a test of the hypothesis that the autoregressive model for X_t contains a unit root against the alternative that this root is less than one.

A *unit root test* can be constructed in the following way. Suppose we have the AR(1) model $X_t = \theta + \phi X_{t-1} + u_t$ and we wish to test the *unit root hypothesis* H_0: $\phi = 1$ against the stationary alternative H_A: $\phi < 1$. An 'obvious' approach would be to estimate the model by OLS and compute the t-statistic $(\hat{\phi}-1)/SE(\hat{\phi})$, rejecting the null if there is a significantly large negative value of the statistic. The problem is that this statistic is *not* distributed as $t(T-2)$ on H_0, as might be expected from following the analysis of §12.3. The distribution that the statistic follows is known as the *Dickey–Fuller distribution*, is usually denoted τ_μ to distinguish it from t, and is shown, with the standard normal for comparison, in Figure 18.9.[15]

For large samples, the 5%, 2.5% and 1% critical values of τ_μ are −2.86, −3.12 and −3.43, as compared to −1.645, −1.96 and −2.33 for the standard normal, so that incorrectly using these latter critical values substantially increases the probability of making a Type 1 error. For example, using −1.645 rather than the correct −2.86 as the critical value will increase the probability of incorrectly rejecting a unit root null in favour of the stationary alternative from 5% to over 46%. Since the AR(1) model can be written as

$$\Delta X_t = \theta + (\phi - 1)X_{t-1} + u_t = \theta + \varphi X_{t-1} + u_t \tag{18.7}$$

the test statistic τ_μ may be obtained directly as the t-ratio on $\hat{\varphi} = \hat{\phi} - 1$, the slope coefficient in the regression of ΔX_t on X_{t-1}.[16]

Higher order autoregressions may be tested for the presence of a unit root by a straightforward extension of these ideas. Suppose we consider the AR(2) process, for which the unit root null hypothesis is $\phi_1 + \phi_2 = 1$. Analogous to (18.7), an AR(2) model can be written as

$$\Delta X_t = \theta + (\phi_1 + \phi_2 - 1)X_{t-1} - \phi_2 \Delta X_{t-1} + u_t \tag{18.8}$$

Thus the unit root null can be tested using the t-ratio of X_{t-1} in the regression of ΔX_t on X_{t-1} and ΔX_{t-1}, and this will continue to follow the τ_μ distribution. Generally, a unit root in an AR(p) process can be tested using an analogous regression to (18.8) but with p–1 lags of ΔX_t included as additional regressors. Such a regression is known as an *augmented Dickey–Fuller (ADF) regression*, and the associated test an *ADF test*. Typically the order of the autoregression will be unknown, and p must be selected by, say, an information criterion or some other method (cf. §**18.3**).

Testing for unit roots in the spread and the exchange rate

The AR(2) specification for the spread can be written as the ADF regression

$$\underset{}{\Delta X_t} = \underset{(0.018)}{0.034} - \underset{(0.008)}{0.032}\, X_{t-1} + \underset{(0.036)}{0.223}\, \Delta X_{t-1} + \hat{u}_t$$

so that $\tau_\mu = -0.03208/0.22291 = -3.89$, which is *less than* the 1% critical value and thus *rejects* the null that the spread contains a unit root.

Thus, even though the spread contains a very large autoregressive root, this root is significantly less than unity and the spread is stationary around a constant mean. On average, therefore, long rates are just over one percentage point higher than short rates, but the rate spread can lie above or below this equilibrium value for considerable periods of time.[17]

The ADF regression for the exchange rate is

$$\Delta X_t = \underset{(0.015)}{0.039} - \underset{(0.008)}{0.023}\, X_{t-1} + \underset{(0.046)}{0.134}\, \Delta X_{t-1} + \hat{u}_t$$

so that $\tau_\mu = -0.02340/0.00835 = -2.80$. Since this is *greater than* the 5% critical value, the hypothesis that the exchange rate contains a unit root cannot be rejected at this significance level – although, interestingly, it can be rejected at the 10% level (the p-value is 0.0565). Thus the exchange rate appears to follow a driftless but slightly autocorrelated random walk: note that conventional but incorrect inference would have clearly, but erroneously, rejected a unit root in favour of stationarity around a constant equilibrium exchange rate!

18.8 Trend stationarity versus difference stationarity

§6.6 also introduced the time trend regression $X_t = \beta_0 + \beta_1 t + e_t$, which may be thought of as another way of transforming a non-stationary X_t to stationarity if the *detrended* $e_t = X_t - \beta_0 - \beta_1 t$ is itself stationary.

Whether this is the case or not can be investigated using an extension of the Dickey–Fuller test. The time trend regression can be considered to be the *trend stationary* (TS) alternative to the *difference stationary* (DS) null $X_t = X_{t-1} + \theta + e_t$, that is, the DS null states that X_t is generated by a (possibly autocorrelated) drifting random walk, while the TS alternative is that X_t is generated as stationary fluctuations around a linear trend.[18]

Formally, consider the following 'structural' model, in which X_t is generated as a linear trend 'buried' in noise, where an autoregression of order two for e_t is assumed for expositional convenience

$$X_t = \beta_0 + \beta_1 t + e_t \tag{18.9}$$

$$e_t = \phi_1 e_{t-1} + \phi_2 e_{t-2} + u_t \tag{18.10}$$

Substituting (18.9) into (18.10)

$$X_t - \beta_0 - \beta_1 t = \phi_1 \left(X_{t-1} - \beta_0 - \beta_1 (t-1)\right) + \phi_2 \left(X_{t-2} - \beta_0 - \beta_1 (t-2)\right) + u_t$$

and rearranging gives

$$\begin{aligned} \Delta X_t = & \left(-\beta_0\left(\phi_1 + \phi_2 - 1\right) + \beta_1\left(\phi_1 + 2\phi_2\right)\right) - \beta_1\left(\phi_1 + \phi_2 - 1\right) t \\ & + \left(\phi_1 + \phi_2 - 1\right) X_{t-1} - \phi_2 \Delta X_{t-1} + u_t \end{aligned} \tag{18.11}$$

If X_t contains a unit root then $\phi_1 + \phi_2 = 1$ and (18.11) reduces to

$$\Delta X_t = \beta_1(1+\phi_2) - \phi_2 \Delta X_{t-1} + u_t$$

or

$$\Delta X_t - \beta_1 = -\phi_2(\Delta X_{t-1} - \beta) + u_t$$

that is, the structural model

$$X_t = X_{t-1} + \beta_1 + e_t^*$$

$$e_t^* = -\phi_2 e_{t-1}^* + u_t$$

where $e_t^* = \Delta e_t$. A test of the unit root null is thus given by the t-ratio on X_{t-1} in (18.11). More generally, if an AR(p) is specified in (18.10), then (18.11) can be written as the extended ADF regression

$$\Delta X_t = \theta + \beta t + \varphi X_{t-1} + \sum_{i=1}^{p-1} \phi_i \Delta X_{t-i} + u_t \qquad (18.12)$$

where θ and β are functions of β_0, β_1 and the ϕ_i and $\varphi = \phi_1 + \phi_2 + \ldots + \phi_{p-1}$. The DS null hypothesis is thus H_0: $\varphi = 0$ and the TS alternative is H_A: $\varphi < 0$ and the test statistic is again the t-ratio on $\hat{\varphi}$. The presence of t as a regressor in the ADF regression, however, alters the distribution of the test statistic to that shown in Figure 18.10, and the statistic is denoted τ_τ accordingly. The large T 5%, 2.5% and 1% critical values of τ_τ are −3.41, −3.66 and −3.96, so that the presence of a drift term in the null model requires an even more extreme test statistic for the null to be rejected.

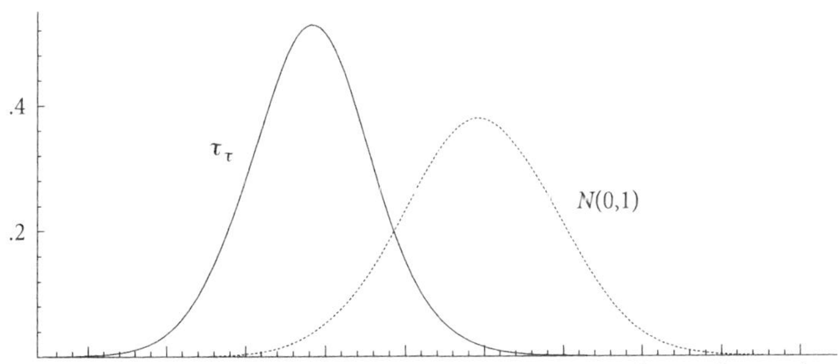

FIGURE 18.10 *The Dickey–Fuller τ_μ distribution compared to the standard normal*

Are consumption and income trend or difference stationary?

The time trend regressions for consumption and income reported in §6.6 clearly assume that their logarithms were generated as TS processes– but is this in fact the case? Estimates of (18.12) with $p = 2$, the order chosen by the *BIC*, for the two series, are

$$\underset{}{\Delta \ln C_t} = \underset{(0.704)}{2.442} + \underset{(0.0014)}{0.0047}\, t - \underset{(0.056)}{0.194} \ln C_{t-1} + \underset{(0.111)}{0.546}\, \Delta \ln C_{t-1} + \hat{u}_t$$

$$\Delta \ln Y_t = \underset{(0.899)}{2.943} + \underset{(0.0018)}{0.0056}\, t - \underset{(0.071)}{0.231} \ln C_{t-1} + \underset{(0.122)}{0.456}\, \Delta \ln Y_{t-1} + \hat{u}_t$$

From these regressions, the τ_τ statistics for consumption and income may be calculated to be $-0.1943/0.0563 = -3.45$ and $-0.2313/0.0711 = -3.25$ respectively. Since the 10% and 5% critical values for this sample size are -3.17 and -3.49, both statistics reject the DS null at the 10% level but not at the 5% level, the *p*-values being 0.0544 and 0.0839 respectively. This confers some doubt as to the correct specification for both series, particularly as Dickey–Fuller tests can suffer from rather low power (recall §11.5).

Consequently, it seems worth reporting both types of model. The structural TS models are

$$\ln C_t = \underset{(0.02)}{12.51} + \underset{(0.0005)}{0.0243}\, t + e_t$$

$$e_t = \underset{(0.112)}{1.352}\, e_{t-1} - \underset{(0.111)}{0.546}\, e_{t-2} + \hat{u}_t \qquad \hat{\sigma} = 1.44\%$$

and

$$\ln Y_t = \underset{(0.02)}{12.67} + \underset{(0.0006)}{0.0244}\, t + e_t$$

$$e_t = \underset{(0.119)}{1.225}\, e_{t-1} - \underset{(0.122)}{0.456}\, e_{t-2} + \hat{u}_t \qquad \hat{\sigma} = 1.78\%$$

while the structural DS models are

$$\Delta \ln C_t = \underset{(0.0036)}{0.0230} + e_t^*$$

$$e_t^* = \underset{(0.117)}{0.451}\, e_{t-1}^* + \hat{u}_t \qquad \hat{\sigma} = 1.56\%$$

and

$$\Delta \ln Y_t = \underset{(0.0037)}{0.0238} + e_t^*$$

$$e_t^* = \underset{(0.123)}{0.335}\, e_{t-1}^* + \hat{u}_t \qquad \hat{\sigma} = 1.90\%$$

The TS structural models can be thought of as 'decomposing' the observed series into additive (linear) trend and cyclical (AR(2)) components (cf. §6.6; also recall the decompositions of §3.5), as shown graphically in Figure 18.11.

The trend lines are essentially parallel since trend growth is almost identical for both series at approximately 2.43%. The AR(2) representations of the cyclical components both have a pair of complex roots, $0.68 \pm 0.30i$ for consumption and $0.61 \pm 0.28i$ for income: these imply that the components follow 'stochastic' sine waves with average periods of 15.1 and 14.5 years respectively, the nature of which are clearly seen in the figure.[19]

The DS structural models show that both series follow an AR(1) autocorrelated random walk with drift (essentially ARIMA(1,1,0) processes). There is no unique trend-cycle decomposition of such a model, but the *Beveridge-Nelson decomposition* is popular. Here $X_t = T_t + I_t$ with the trend, T_t, and cycle, I_t, components being *assumed* to be perfectly correlated with each other and given by[20]

$$T_t = T_{t-1} + \beta_1 + \frac{1}{(1+\phi_2)} u_t$$

$$I_t = \phi_2 I_{t-1} + \frac{\phi_2}{(1+\phi_2)} u_t$$

Thus the trend is a random walk with the same drift as X_t and the cycle is a (stationary) AR(1) process. The components are perfectly correlated because they are 'driven' by the same innovation u_t. Thus, for consumption,

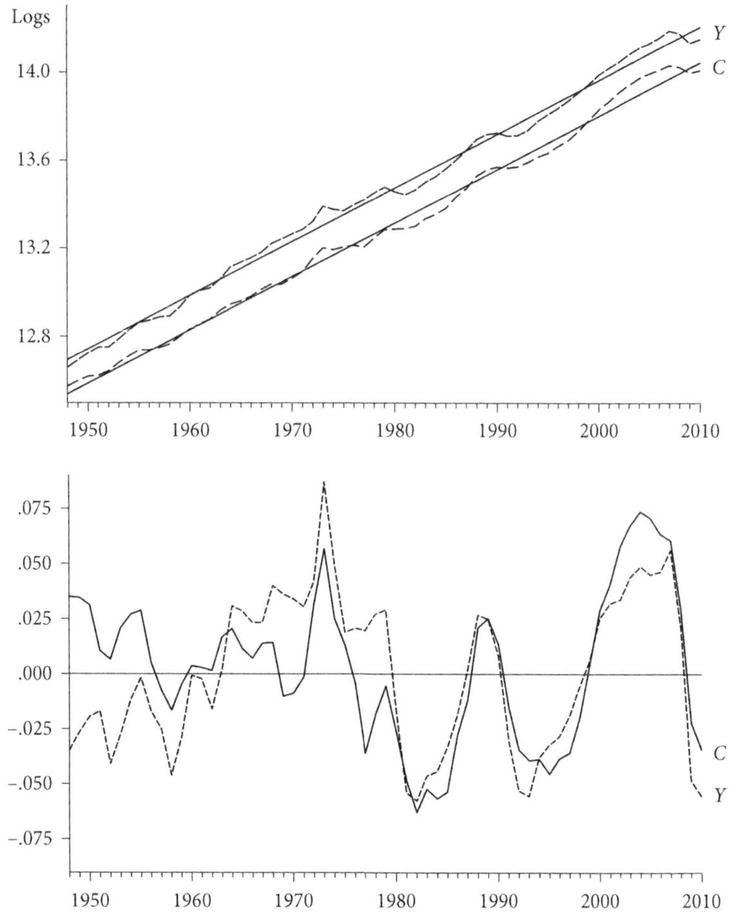

FIGURE 18.11 *Logarithms of consumption and income with fitted trends (top panel) and cyclical components (bottom panel) from TS models*

$$T_t = T_{t-1} + 0.0230 + 1.821u_t, \qquad I_t = -0.451I_{t-1} - 0.821u_t$$

while for income

$$T_t = T_{t-1} + 0.0238 + 1.504u_t. \qquad I_t = -0.335I_{t-1} - 0.504u_t$$

With these decompositions, consumption and income have trend growth rates that are approximately equal *on average*, but the trends evolve as drifting random walks rather than as deterministic linear functions. The

AR(1) models for the cyclical components preclude any business cyclical interpretation of their fluctuations.

TS and DS models have a further important difference. As we have seen, the TS model is $X_t = \beta_0 + \beta_1 t + e_t$, where e_t is *stationary*. For example, if $e_t = \phi e_{t-1} + u$, then

$$X_t = \beta_0 + \beta_1 t + \phi^2 e_{t-2} + u_t + \phi u_{t-1}$$
$$\vdots$$
$$X_t = \beta_0 + \beta_1 t + u_t + \phi u_{t-1} + \ldots + \phi^k u_{t-k} + \ldots$$

using the stationarity condition $|\phi| < 1$. Thus the effect of *any* innovation on X_t must dissipate through time, and the series has to revert to its unique trend line, so that *shocks are transitory.*

On the other hand, the DS model $X_t = X_{t-1} + \beta_1 + e_t^*$, where e_t^* is stationary, can be written as

$$X_t = X_{t-2} + 2\beta_1 + e_t^* + e_{t-1}^*$$
$$\vdots$$
$$X_t = X_0 + \beta_1 t + e_t^* + e_{t-1}^* + \ldots + e_1^*$$

that is, as a linear trend but with a *non-stationary* error, this being the accumulation of all past innovations. Since any innovation e_{t-k}^* remains unweighted in this accumulation, its effect on X_t fails to die away so that *shocks are permanent.*

This difference has the implication that if consumption and income were TS, then a major recessionary shock, such as that of 2008, will have only a temporary effect so that after a period of negative growth, consumption and income must subsequently grow more rapidly than 'normal' to return to the trend path. If these series are DS, however, such a major recessionary shock remains in the data for ever, and there is *no* tendency for the series to return to any unique trend path: in fact, the trend paths alter each period in response to the latest innovation.

18.9 Further models and tests

This chapter has only scratched the surface of time series modelling. Seasonal time series, such as retail sales analysed in §3.5, can be modelled using an extension to seasonal ARIMA models. Several time series may be analysed jointly by using *vector* extensions of the models introduced here. These models can also be used to investigate heteroskedasticity that changes through time, producing the class of *generalised autoregressive conditional heteroskedastic* (GARCH) models that have become the standard way of modelling *volatility* in financial time series data.

A number of variants to the Dickey–Fuller unit root test have been proposed, essentially to improve the power properties of such tests, so that a clear distinction between the TS and DS models for consumption and income might be established. These tests have also been extended to cover the cases of potentially multiple unit roots, breaks in the series, and situations in which combinations of integrated time series may be stationary, the phenomenon known as *co-integration*.[21]

Notes

1 Autoregressions were first introduced by the famous British statistician George Udny Yule during the 1920s: see G.U. Yule, 'On a method of investigating periodicities in disturbed series, with special reference to Wolfer's sunspot numbers', *Philosophical Transactions of the Royal Society of London, Series A* 226 (1927), 267–298. For details on the historical development of these models, see Terence C. Mills, *The Foundations of Modern Time Series Analysis* (Palgrave Macmillan, 2011) and *A Very British Affair: Six Britons and the Development of Time Series Analysis* (Palgrave Macmillan, 2013). Acronyms abound in time series analysis and have even prompted a journal article on them: Clive W.J. Granger, 'Acronyms in time series analysis (ATSA)', *Journal of Time Series Analysis* 3 (1982), 103–107, although in the three decades since its publication many more have been suggested.

2 The term 'white noise' was coined by physicists and engineers because of its resemblance, when examined in the 'frequency domain', to the optical spectrum of white light, which consists of very narrow lines close together: see Gwilym M. Jenkins, 'General considerations in the analysis of spectra', *Technometrics* 3 (1961), 133–166. The term 'innovation' reflects the fact that the current error u_t is, by definition, independent of all previous values of both the error and X and hence represents unforecastable 'news' becoming available at time t.

3 The random walk $X_t = a + X_{t-1} + u_t$ can be written, on successively substituting for lagged X's back to the initial value X_0, as $X_t = X_0 + ta + u_t + u_{t-1} + \ldots + u_1$,

from which the expressions for the mean and variance of X_t are obtained immediately.

4 Writing (18.1) in the form of (18.3), that is, as $(X_t - \mu) = b(X_{t-1} - \mu) + u_t$, then

$$\begin{aligned} V(X_t) &= E(X_t - \mu)^2 = E\left(b(X_{t-1} - \mu) + u_t\right)^2 = b^2 E(X_{t-1} - \mu)^2 + u_t^2 \\ &= b^2 V(X_t) + \sigma^2 \\ &= \sigma^2 / (1 - b^2) \end{aligned}$$

Note that the stationarity condition $|b| < 1$ also ensures that $V(X_t)$ is positive and finite.

5 The successive sample partial autocorrelations may be estimated recursively using the updating equations proposed by James Durbin, 'The fitting of time series models', *Review of the International Statistical Institute*, 28 (1960), 233–244, which are known as the *Durbin–Levinson algorithm*:

$$\hat{\phi}_{kk} = \frac{r_k - \sum_{j=1}^{k-1} \hat{\phi}_{k-1,j} r_{k-j}}{1 - \sum_{j=1}^{k-1} \hat{\phi}_{k-1,j} r_j}$$

$$\hat{\phi}_{kj} = \hat{\phi}_{k-1,j} - \hat{\phi}_{kk} \hat{\phi}_{k-1,k-j} \qquad j = 1, 2, \ldots, k-1$$

See George E.P. Box and Gwilym M. Jenkins, *Time Series Analysis: Forecasting and Control*, revised edition (Holden Day, 1976), pages 177–178, for discussion of the standard errors to be attached to both sample autocorrelations and sample partial autocorrelations.

6 These information criteria may be used to select between any competing regression models explaining the same dependent variable, with k interpreted as the number of parameters fitted in a particular model.

7 This is also known as the *Ljung–Box statistic*: Greta M. Ljung and George E.P. Box, 'On a measure of lack of fit in time series models', *Biometrika* 65 (1978), 297–303.

8 This result may be obtained directly on noting that

$$\begin{aligned} Cov(X_t, X_{t-k}) &= E(X_t - \mu)(X_{t-k} - \mu) \\ &= E\left(b^k (X_{t-k} - \mu) + u_t + u_{t-1} + \ldots + u_{t-k+1}\right)(X_{t-k} - \mu) \\ &= b^k E(X_t - \mu)(X_{t-k} - \mu) = b^k V(X_{t-k}) = b^k V(X_t) \end{aligned}$$

9 It is clear that for the MA(1) process, $\mu = \theta$. Thus

$$\begin{aligned} Cov(X_t, X_{t-k}) &= E(X_t - \theta)(X_{t-k} - \theta) \\ &= E\left(u_t u_{t-k} + \vartheta u_t u_{t-k-1} + \vartheta u_{t-1} u_{t-k} + \vartheta^2 u_{t-1} u_{t-k-1}\right) \end{aligned}$$

For $k = 0$, $V(X_t) = \sigma^2(1 + \vartheta^2)$, for $k = 1$, $Cov(X_t, X_{t-1}) = \sigma^2 \vartheta$, and for $k > 1$, $Cov(X_t, X_{t-k}) = 0$.

10 From Box and Jenkins, *op. cit.*, p. 70, we have

$$\phi_{kk} = \frac{-(-\vartheta)^k\left(1-\vartheta^2\right)}{1-\vartheta^{2(k+1)}}$$

Thus $|\phi_{kk}| < \vartheta^k$ and the PACF is dominated by a geometric decline. If ρ_1 is positive, so that ϑ is positive, the partial autocorrelations alternate in sign, whereas if the converse holds the partial autocorrelations will all be negative.

11 Estimation of models with moving average errors is usually carried out by *conditional* least squares (CLS), where the initial values of the error series that are required for estimation are set to their conditional expectation of zero.

12 The algebraic derivation of this result is as follows. First write the process as

$$X_t - \mu = \phi\left(X_{t-1} - \mu\right) + u_t + \vartheta u_{t-1}$$

where $\mu = \theta/(1-\phi)$. Next obtain $V(X_t)$, $Cov(X_t, X_{t-1})$ and $Cov(X_t, X_{t-k})$ for $k > 1$:

$$\begin{aligned} V(X_t) &= E\left(X_t - \mu\right)^2 = E\left(\phi\left(X_{t-1} - \mu\right) + u_t + \vartheta u_{t-1}\right)^2 \\ &= \phi^2 V(X_t) + \sigma^2\left(1 + \vartheta^2 + 2\phi\vartheta\right) \\ &= \sigma^2 \frac{1+\vartheta^2+2\phi\vartheta}{1-\phi^2} \end{aligned}$$

$$\begin{aligned} Cov(X_t, X_{t-1}) &= E\left(X_t - \mu\right)\left(X_{t-1} - \mu\right) = E\left(\varphi\left(X_{t-1} - \mu\right) + u_t + \vartheta u_{t-1}\right)\left(X_{t-1} - \mu\right) \\ &= \varphi V(X_t) + \sigma^2\vartheta \\ &= \sigma^2 \frac{\left(1+\varphi\vartheta\right)\left(\varphi+\vartheta\right)}{1-\varphi^2} \end{aligned}$$

$$\begin{aligned} Cov(X_t, X_{t-k}) &= E\left(X_t - \mu\right)\left(X_{t-k} - \mu\right) = E\left(\phi\left(X_{t-1} - \mu\right) + u_t + \vartheta u_{t-1}\right)\left(X_{t-k} - \mu\right) \\ &= \phi Cov\left(X_t, X_{t-k+1}\right) \end{aligned}$$

Using these expressions the results for ρ_1 and ρ_k, $k > 1$, follow automatically.

13 The inclusion of dummy variables into an ARMA specification produces what has become known as an *intervention model* in the time series literature: see George E.P. Box and George C. Tiao, 'Intervention analysis with application to economic and environmental problems', *Journal of the American Statistical Association* 70 (1975), 70–79.

14 For an AR(2) process, it can be shown that

$$A_1 = \frac{g_1\left(1-g_2^2\right)}{\left(g_1-g_2\right)\left(1+g_1g_2\right)} \qquad A_2 = \frac{g_2\left(1-g_1^2\right)}{\left(g_2-g_1\right)\left(1+g_1g_2\right)}$$

Thus if $g_1 \approx 1$ and $g_2 \approx 0$, then $A_1 \approx 1$.

15 The seminal article on what has become a vast subject, and which gives the distribution its eponymous name, is David A. Dickey and Wayne A. Fuller, 'Distribution of the estimators for autoregressive time series with a unit root', *Journal of the American Statistical Association* 74 (1979), 427–431. The statistical theory underlying the distribution is too advanced to be considered here but see, for example, Kerry Patterson, *A Primer for Unit Root Testing* (Palgrave Macmillan, 2010) and, at a rather more technical level, his *Unit Root Tests in Time Series. Volume 1: Key Concepts and Problems* (Palgrave Macmillan, 2011).

16 Strictly, the τ_μ statistic tests $\phi = 1$ *conditional* upon $\theta = 0$, so that the model under H_0 is the driftless random walk $X_t = X_{t-1} + u_t$. The joint hypothesis $\theta = 0$, $\phi = 1$ may be tested by constructing the usual F-statistic along the lines of §13.3, although clearly the statistic will not follow the $F(2, T-2)$ distribution. For large samples the 5% and 1% critical values of the appropriate distribution are 4.59 and 6.53, rather than the 2.99 and 4.60 critical values of the F-distribution.

17 We are also now able to confirm the appropriateness of imposing a unit root on the interest rate pre-1965 in §17.5, for here $\tau_\mu = (1-0.956)/0.024 = -1.84$. As this is insignificant, we cannot reject the hypothesis that the interest rate is indeed a random walk.

18 The TS and DS terminology was introduced by Charles R. Nelson and Charles I. Plosser, 'Trends and random walks in macroeconomic time series', *Journal of Monetary Economics* 10 (1982), 139–162.

19 The period, f, of the cycle can be obtained by solving the equation

$$f^{-1} = \cos^{-1}\left(|\phi_1| / 2\sqrt{-\phi_2}\right)/360$$

See Box and Jenkins, *op. cit.*, pp. 58–63.

20 Stephen Beveridge and Charles R. Nelson, 'A new approach to decomposition of economic time series into permanent and transitory components with particular attention to measurement of the "business cycle"', *Journal of Monetary Economics* 7 (1981), 151–174. For a recent introduction to models of this type, see Terence C. Mills, 'Trends, cycles and structural breaks', in Nigar Hashimzade and Michael A Thornton (editors), *Handbook on Empirical Macroeconomics* (Edward Elgar, 2013).

21 These extensions have been covered in, for example, Terence C. Mills and Raphael N. Markellos, *The Econometric Modelling of Financial Time Series*, 3rd edition (Cambridge University Press, 2008).

Index

9781137401892